Become a U.S. Commercial Drone Pilot

John D. Deans

Self-Counsel Press
(a division of)
International Self-Counsel Press Ltd.
USA Canada

Printed in Canada

First edition: 2016

ISBN: 978-1-77040-268-3

Library of Congress Control Number: 2015959531

Self-Counsel Press Inc.
(a subsidiary of)
International Self-Counsel Press Ltd.

Bellingham, WA
USA

North Vancouver, BC
Canada

Contents

Samples

Notice to Readers

Laws are constantly changing. Every effort is made to keep this publication as current as possible. However, the author, the publisher, and the vendor of this book make no representations or warranties regarding the outcome or the use to which the information in this book is put and are not assuming any liability for any claims, losses, or damages arising out of the use of this book. The reader should not rely on the author or the publisher of this book for any professional advice. Please be sure that you have the most recent edition.

Acknowledgments

After leaving the urban, corporate world to become a rural computer consultant in 1998, my wonderful wife Beth put up with a huge transition those first few years. Now, 17 years later, I am again migrating; out of the IT world into the role of a professional photographer with a flying camera. Thank you to my lovely Beth for your support for this project and many others over the years.

I also want to thank my two teenage daughters for their patience when I interrupted their Netflix movies so I could watch the latest aerial on the 65-inch LED Smart TV in full 1080p brilliance. They would say, "That looks great, Dad! But, can you please buy some new background music? This one is getting old."

My then 13-year-old daughter acted as my first observer with her eagle-sharp eyes, helping me keep an eye on the Phantom during many of our initial aerial projects. Thank you, Danielle, for standing by me and keeping our bird out of the trees.

Special thanks to her older sister, Jacqueline, who proofread this book during the writing. Every day a hard copy of the latest chapter was waiting for her to correct with a red pen when she drove home from school.

I'm a lucky man to have such smart, beautiful, and supportive ladies in my life; I love you all.

Introduction

After experiencing the computer boom in the early 1980s, I'm seeing the same initial eruption of a new, advanced industry; one that combines aviation, technology, and photography creating the unmanned aerial vehicle (UAV) market. Also known as drones, these airborne technological wonders have exploded onto both consumer and commercial environments creating opportunities for all. As legal entities such as the Federal Aviation Administration (FAA) and state legislatures catch up with the fast-developing industry, marketplace rules will soon be established for a high-flying entrepreneurial race to begin.

The goal of this book is to provide a comprehensive roadmap on how to become a commercial drone pilot and earn a good income creating beautiful aerial videos, 2D photographic mapping, and other UAV-based aerial services. We will focus on the most popular drone platform, the DJI Phantom line, take you through the current FAA UAV licensing processes, and describe in detail how to start and run a UAV-based aerial photography business. The market share for DJI is estimated to hit $1 billion in 2015, so they are the safe bet for the best and smartest UAV available, and they have the capital for good support and future product development in the years to come.

1. Original, Old-school RC

The old name for drones and UAVs is Remote Controlled (RC) model airplanes and helicopters. As a kid in the 1970s, I remember many kids with cool dads who had impressive RC planes, which they built and flew as father and son/daughter bonding projects. Back then it took both true piloting skills and a nearly required background in small gas-engine maintenance to get those little motors started and keep them running properly. There were no technical aides like First Person View (FPV), GPS-guided flight, or a "return home" feature.

RC pilots back then had no choice but to maintain line of sight with the aircraft and be responsible for all aspects of takeoff, flight pattern, and landing it in one piece without knowing the exact amount of fuel left in the tiny tank. Most of these early RC configurations were put together by hand with off-the-shelf components from a local hobby shop, or ordered from a model plane catalog. Painstaking efforts were made and numerous hours spent carefully assembling the airframe, mounting the wings, and connecting all the airfoil control surfaces. This was followed by testing sessions to verify the RC controller was compatible and reliable with the model plane's receiver unit.

After all the workshop labor was completed, it was time to head to the park and try the first test flight. If you were able to finger-start the prop without losing a digit, the time finally came to test your flying skills for real, but while standing on the ground. Remember, this was before computer or Internet-based flight simulators were in the home, so usually only true private pilots or seasoned commercial passenger jet captains were able to smoothly operate their airborne creations.

Lord help you if the funds were available and you built an RC model helicopter. Those who thought flying a fixed-wing RC plane in a pattern and landing smoothly in a field was difficult never attempted flying an RC helicopter; those who did were likely to crash it the very first day. Even after basic flight maneuvers were learned, any small mistakes at low altitude or strong gusts of wind during landing could make a dangerous accident occur quickly. Life and limb were in jeopardy when a two-foot (or larger) radius RC helicopter rotor broke up during a rough landing.

A couple of generations ago, flying those RC crafts took patience, skill, and actual aeronautical knowledge. One was always aware of landing zone options, fuel levels, altitudes, wind direction, airspeed, and visibility.

2. UAVs Are Here

Fast forward a few decades, and we now have an explosion of high-tech quadcopters that can be removed from the box, flown autonomously to GPS waypoints, and viewed on an iPad within minutes of delivery from Amazon.com. By the way, how long will it be until Amazon delivers your new drone with an Amazon delivery UAV to your home an hour after your online purchase, given they're considering drone deliveries?

We now have the low-cost and limited-feature quadcopter drones from Parrot starting at $100 all the way up to the long-range and heavy payload oct-rotor UAVs for $10,000 or more. For ultra-range UAVs, both the consumer and the commercial pilot can opt for fuel-efficient, fixed-wing planes that can fly waypoint courses for tens to hundreds of miles if you are willing to make that investment.

Again, the drone platform we are going to focus on in this book is the most popular UAV on the market, the DJI Phantom 2 Vision Plus and Phantom 3 quadcopters. Tens of thousands of DJI Phantoms have been purchased, flown, crashed, and enjoyed in the US over the past couple of years, and this is just the start. It is one thing to take out your new Phantom, get it a couple of hundred feet in the air above your subdivision and start taking pictures. It is another to plot out a strategy on how to utilize this amazing flying camera in a profit-making venture, during the birth of the UAV commercial market in a barely legal environment (barely legal at the moment, because there is currently a ban on commercial "droning" unless you have an FAA 333 exemption, which we will discuss in Chapter 11).

This book is a comprehensive guide on how to make money with your DJI Phantom in a robust but legal manner. We will cover all aspects of what it takes to develop your aeronautical skill sets, commercial photography capabilities, small-business marketing techniques, safety procedures, video editing processes, end product deliverables, computer software for aerial mapping, and UAV maintenance practices.

3. Why Am I the Right Person to Teach You How to Make Money with Drones?

I was a National Honor Society "A" student at Bellaire High School in Houston, Texas, but I despised college at the University of Houston. So, before the first week of my freshman semester was over, I was out. Since I was only 18 years old, I had to wait a year before I could apply to the Houston Police Academy to start a career in law enforcement.

During that year, my dad found me a job at Control Data Corporation (CDC) as a process control clerk working with computers. This began my 35+ year career in Information Technology (IT), as a fluke; I never joined HPD. The simple reason was I was making more than the cops were when I became a programmer in less than a year.

The Author and His Plane

During that same year in 1982, I began flight school to get my private pilot's license flying out of Houston Hobby airport. I was trained on a low-wing Grumman Cheetah, and before the year was over I passed my check ride and became an FAA-licensed private pilot.

Those were the simpler days. I worked hard in the computer rooms to earn plane rental money for the weekend. Since there were no digital cameras back in the 1980s and I could not afford a big-dollar SLR camera, I'd fly down the Galveston beach with the plane's canopy pulled all the way back, while holding my Kodak Instamatic film camera with a dozen exposures trying to get a good aerial shot from only a few hundred feet above the surf. Try to fly that low nowadays and you'd probably have at least the FAA on you when you landed, if not the National Guard out of Ellington military base on the trip back.

After three years, I left CDC as a systems analyst to migrate oil field reservoir simulation software to every type of super computer available from 1984 to 1987, while working with J. S. Nolen and Associates. During that time I married my first wife and had a son. Aerial anything went to the very back burner. I guess that is what happens when the responsibility of a family hits you when you are young.

Things got interesting again when I became manager of computer operations at CogniSeis Development and started making some good money in 1988. The downside of those days is my marriage broke up, but the upside was I got to start flying again and spend some good one-on-one time with my growing boy.

When the first digital cameras came out in the early 1990s, I had to try them out by flying over downtown Houston. Remember, this was pre-9/11, so it was no big deal to fly near buildings at less than 1,000 feet Above Ground Level (AGL). By that time, I was a networking consultant being billed out by Paranet to Fortune 100 companies in Houston like Amoco, British Petroleum, and Compaq Computer. I borrowed the company's first digital camera, which cost more than $1,000 and only took black and white images.

My boss wanted to take some aerial pictures of his home in Memorial, so we used the new corporate camera. The neat thing about the digital camera was that I could see my aerial photos as soon as I got home and downloaded them to my Windows 3.1 desktop, for which I paid more than $3,000 back in 1995.

My first semiprofessional aerial photography shoot was a bust since his house was completely surrounded by trees, and I refused to fly the rented Cheetah at his requested low altitude of 200 feet above the rooftops. That was more than 20 years ago, but I remember thinking how cool it would be if I could mount this bulky digital camera on a model airplane to have a "flying camera." I could fly it from the ground to take pictures at very low altitudes with little risk to myself and other people.

I married my wonderful wife Beth in 1995 and we now have two beautiful teenage daughters. Since their births I have taken thousands of pictures with my Nikon D5000 SLR. Each vacation destination has inspired me to build up my lens variety, try different scene techniques, and constantly learn more about photography. My images got even clearer after I took photography classes and sought photographic tutoring from local professionals.

After 17 years in the high-stress world of IT while fighting the hellish traffic in Houston on a daily basis, I cashed out my Paranet stock and left the corporate world in 1998. We purchased a 115-acre ranch just west of Brenham, Texas, and I started Deans Consulting in 2000. During my attempted escape from the computer industry, I looked at every agri-business from raising chickens for eggs and meat, goats for meat, sheep for wool, horses for boarding, grapes for wine, and numerous other ventures. None of these panned out and I quickly discovered that I could not make near the same money on my small ranch as I could rejoining the IT world in the business community of a small town.

Deans Consultant, a.k.a., The Country Computer Consultant, has been highly profitable for more than 16 years, and odds are I'll have a hand in it for years to come, since I love my clients and it's relatively easy IT work. My self-employment has only caused a fraction of the stress I felt in the 1990s, since there is no plane travel, minimal traffic, no contracts, and really nice country people to work for here in Washington County.

However, being in my mid-50s, crawling under desks to find data cabling problems, climbing ladders to mount wireless access points, and having the heavy responsibility of data integrity and security for numerous small companies has been weighing on me over the past few years.

In 2014, I starting a part-time career as a firearms instructor teaching Concealed Handgun Licensing (CHL) classes at the gun range on our ranch. The next year I developed numerous gun fighting simulation services for law enforcement and CHL clients. I was making money outside of IT, but I was still looking for something else to make a second attempt of finding a new career outside of my 35-year IT identity.

That same year, a new era of aerial photography started with the availability of the DJI Phantom 2 Vision Plus. That changed everything for me. After seeing incredible aerial videos and still photos on YouTube and the Internet, I had to purchase one for at least testing purposes. This was the "flying camera" I dreamed about 20 years earlier, and the possibilities for new business opportunities went off like fireworks in my head.

From mid-2014 and through 2015 I flew more than 100 test flights at buddies' ranches and IT clients' construction sites without charging for the flights or the videos I produced. My goal was to explore these test markets in a "hobby" mode to see if a real business could be derived from flying drones.

As it turned out, this was it! This was my ticket out of IT and into a new, booming industry that is exciting, challenging, beautiful, and potentially very profitable. I'm in! I'm all in!

The Author, John D. Deans

1 Is Commercial "Droning" Right for You?

Flying UAVs in a hobby mode at the park or filming your buddy's farm for fun is one thing, but to start a small business based on uncharted market waters, with highly technical apparatus in a confusing legal matrix of restrictions, and make money on the whole endeavor is quite another world.

First and foremost, you will need the passion for it. As with any other career path, to be successful you need to have that fire in the belly; the yearning to get up every day early to check the weather. Wanting to fly and earn must be there naturally. Even if you have the best UAV configuration, hold all the legal commercial pilot credentials, and have customers who want your aerial photography services, you also need to have a small-business mentality.

1. Starting a Business and Wearing Many Hats

It is difficult for most people to resign from stable, full-time jobs with consistent paychecks and medical insurance benefits, to start up new jobs based on a brand new marketplace and be able to successfully plan, execute, and survive past the first few years. Small-business owners must have not only the desire to be independent and prosperous, but also possess the skill sets to wear the numerous hats required to start a new company.

Those roles include business plan developer, service/product designer, salesperson, bookkeeper, customer service representative, bill collector, tax preparer, photographer, software and hardware technician, moviemaker, and finally, drone/UAV pilot. Those are a lot of hats to wear, and it may take many years to acquire critical business skills and technical capabilities.

I am not trying to scare you away from this fascinating field, but you do need to know what you are contemplating getting into, since most new, small companies fail within the first five years of operation, mostly due to lack of funding and poor preparation. The goal of this book is to help you succeed as a commercial drone pilot, but at the same time you need to understand and face the realities of the free enterprise market. You must be armed with a strong capitalistic attitude and have success in your soul.

2. Your Time

One of the most critical resources for you to manage as a commercial drone pilot is time. Odds are you will not be able to immediately quit your job or leave school to devote 100 percent of your time to droning. Most aerial data acquisitioning projects, including mapping, real estate photography, and construction status aerial surveys are time sensitive. As we will discuss in later chapters in detail, time windows can be critical; you will need to have your bird (what we often call our drones) in the air capturing exactly what your client wants, when he or she wants it captured.

Throw in the massive variable of weather and your time opportunities become much more complex. For example, you may have an agreement with a real estate agent to perform aerial photography on a 50-acre ranch that is for sale, but the agent requested it be filmed in the late afternoon with the golden setting sunlight, and you need to get it done this week, before the farmer moves the hay bales from the meadow. If the sun sets at 6:00 p.m. and you cannot get off work until 5:30 p.m., that can be difficult. There may be high winds or rain several days in a row that hamper your scheduled flights, and time is ticking.

Another example: Your construction client wants you to aerial-film a concrete pour at sunrise. What is the dew and moisture situation?

Can you make it to that location at that time? Do you have work or school commitments and conflicts?

How about filming outdoor weddings and other events? Since they are mostly on the weekend, are you ready to work during the week and on the weekends? What about public holidays? Then there are possible law enforcement opportunities, for which you may be required to be on call 24/7 and be immediately available to perform aerial surveillance over a quickly evolving event. Can you move fast and have your bird ready to fly at a moment's notice?

All these time-related issues need to be contemplated and evaluated. Some aerial markets are more time-sensitive and -consuming than others. When you are just starting up you will be tempted to take any and all aerial opportunities to fly and earn, but be aware of the golden resource of your time, that you must efficiently manage.

3. Location of Markets

What if you have significant free time on your hands; are you in the right place near the right aerial markets? There are market limitations for numerous reasons in the middle of large metropolitan areas such as New York, Chicago, Dallas, and others mainly due to FAA airport area UAV restrictions and local ordinance bans on drones, not to mention line of sight and signal interference problems in downtown areas. There may be tons of people with carloads of money in big cities, but legal and realistic aerial photography opportunities are limited.

The same thing goes for the opposite end of that spectrum in sparsely populated areas like West Texas, where you have ample sky in which to hover and open land to fly over, but few people are around to pay you for it. If there is limited growth in a region then there will be fewer construction projects to film. If land is not selling such as smaller acreage ranches, then fewer real estate agents will be in need of your UAV photography services.

Optimally, you need to be based in a 100-mile radius of economic growth including construction projects, rural developments, industrial activity, and agricultural businesses. The closer you are to multiple aerial market prospects, the more opportunities you will have to sell your aerial photography services.

Also critical to the location you choose will be the state laws and local ordinances restricting UAV activities, which can put a chokehold on your business with the stroke of a bureaucrat's pen, or worse yet, a politician's ill-thought vote.

After you consider all these factors, take a breath and do the math. Can you manage your time well enough in a small-business person's world in the right area of the country to make real money with your flying camera? If your answer is yes, then keep reading. I'll show you how!

2
UAV Restrictions: Federal, State, and Local

Ever since 2014, it seems every news cycle has a story about drones. Whether about somebody flying in a restricted area, a UAV crash-landing somewhere, or a new market use for drones, I was always hearing something about them. Being on the road to becoming a commercial drone pilot and business owner, I had to know more about this fast-developing technology and the bureaucracies that wanted to regulate them. To stay even better informed, I configured a Google Alert so all news articles from around the world with any of the keywords "UAV," "UAS," and "Drone" in the title or content would be emailed to me every morning at 8:00 a.m. This helped me keep up on the new businesses getting granted the FAA 333 exemption, (we'll cover how I got mine in Chapter 11), along with the debate raging in municipalities, state capitals, and congress over drone legislation and restrictions.

FAA Logo

1. The Feds

At the time of the writing of this book, any American could order a quadcopter UAV from Amazon.com during the week and have it airborne that weekend without any governmental approval, as it should be. We could film our homes, photograph the ocean from above, and make a 2D map of terrain from that day in high resolution. This is light years ahead compared to using a Google Earth image that is grainy and more than one year old. But try to sell those services to the general public and you are committing a federal crime with heavy civil penalties of up to $10,000 per offense. No matter where you live in the US, what your state statutes are, or how strict your city ordinances are, the 800-pound gorilla waiting to pounce has always been the FAA.

The possibility of being fined all started in October of 2011 when a pioneer commercial drone pilot named Raphael Pirker flew a fixed-wing Zephyr inside a stadium full of spectators. He was paid to perform this aerial photography by the University of Virginia in Charlottesville. That action broke the FAA's rule against commercial drone activity and landed him a $10,000 fine. In March of 2014, Judge Patrick Geraghty of the National Transportation Safety Board reversed that fine, stating there was "no enforceable FAA rule" or regulation that applied to a model aircraft such as the one Pirker was flying.

That ruling inspired thousands of other UAV pilots to start charging for their aerial services since this reversal was seen as a green light pushing back the FAA. However, just a few months later in November of 2014, a five-member panel of the National Transportation Safety Board reversed the reversal and the ban was back in effect. Finally, in January of 2015, Pirker settled his case for a reduced fine of $1,100 with a statement of "No Wrongdoing."[1] Since then, hundreds of cease and desist letters from the FAA have gone out to start-up commercial drone entities, which scared them enough to at least temporarily ground their UAVs. Frustrated by the FAA threats of $10,000 fines, UAV entrepreneurs started seriously investigating the efforts in attaining the coveted 333 exemption from the commercial drone ban. In Chapter 11, we will cover the steps I took to successfully attain my FAA 333 exemption for my company, Central Texas Drones. There were several other federal hoops I had to jump through before I legally got into commercial drone piloting. These extra steps included renewing my private pilot's license, getting my Third Class Medical Certificate, and

1 "FAA, drone operator settle suit over commercial flyover," Katie Pasek, The Knowledge Effect, Thomson Reuters, accessed November 2015. blog.thomsonreuters.com/index.php/faa-drone-operator-settle-suit-over-commercial-flyover

registering both my Phantoms with the FAA. That's right, I had to get actual November tail numbers for my $1,000 drones!

All federal parks, forests, and monuments have banned all UAVs and drone activity. This is really a shame, since some of the most breathtaking vistas and scenes are on public federal land and Americans cannot use flying cameras to photograph them. The same ban applies to all State parks managed by the Army Corps of Engineers, which oversees numerous lakes and dams. We are banned from droning at our Lake Somerville State Park in Texas just seven miles from my home. State-owned and -operated parks are doing the bans too, like here in Texas at our Washington-on-the-Brazos State Historic Site. Funny how state and federal governments want to quickly ban UAVs outright, rather than implement simple common-sense restrictions like not flying over people.

2. UAV State Laws

State laws add another layer of necessary compliance and education by both casual and commercial drone operators. Our Texas UAV law has a primary ban on using a drone for illicit surveillance. In other words, you cannot fly over your neighbor's private property and take pictures; that would be a class C misdemeanor. Put those same illegally taken images on the web for public viewing and the offense rises to a class B misdemeanor. Our Texas legislators went so far as to ban environmental groups from using drones to take aerial photographs of oil and gas pipelines, but removed the class C charge from taking pictures from the sky within 25 miles of the Texas/Mexico border.[2]

Wisconsin and numerous other states ban "weaponizing" drones and consider that a felony. They make law enforcement obtain a warrant before using UAVs to collect evidence in criminal investigations. The vast majority of states, including Massachusetts and Maryland, passed strong UAV bans on invasion of privacy. Colorado and a few other states banned the use of drones to aid hunters.

Since you are more likely to get nailed by local authorities than the Feds, research your state's statutes to make sure you are not in violation when your drone leaves the ground. I keep a hard copy of our Texas HB 912 Drone Law in my flight case, in the event some ignorant cop wants to take my bird because he thinks I'm breaking the law. With the text handy I can ask him to point out the state law he's accusing me of violating. (Yes, I have actually had to do that.)

2 "Chapter 423: Use of Unmanned Aircraft," Goverment Code, accessed November 2015. www.statutes.legis.state.tx.us/Docs/GV/htm/GV.423.htm

3. City Drone Ordinances

City ordinances and metropolitan bans can be the most strict, depending where you live. They can be Draconian and affect both the hobbyist and the professional "droner." Numerous cities such as Austin, Texas, and Green Bay, Wisconsin, have already banned drones at most events, such as parades and open-air concerts. Some larger cities such as New York and Chicago are pushing for city-wide bans of all UAVs with the exception of some law enforcement (LE) uses and special permitting.

Basically, the more densely populated the city, the more rules and the stricter the ordinances. The same goes for UAV restrictions: The bigger the town, the more roadblocks for drone pilots. This is one of the many nice things I enjoy living on a ranch in a rural Texas county with a total population of 30,000: No UAV ordinances!

The bottom line is know the local ordinances, the state laws, and the FAA regulations before you spin those rotors; do your research online, with the FAA, and contact local law enforcement for more information in your area.

3 UAV Products: Why the Focus on DJI?

My goal with this book is to provide my proven and comprehensive roadmap, helping you to eliminate the numerous show stoppers I encountered over the years I spent becoming a commercial drone pilot. To be this specific about how things can be done, I had to zero in on a UAV platform that was both widely available and capable of consistent delivery of spectacular images and video.

The UAV commercial markets we will be discussing are, for the most part, best served by multi-rotor drones, and primarily by cost-effective quadcopters. Some long-range mapping projects may require higher-end fixed-wing UAVs, but you can cover more commercial aerial opportunities with a reliable quadcopter platform with all the necessary accessories and travel components.

1. The Technological Turning Points

In 2013, I began researching UAVs for future business prospects, since a real estate buddy of mine was “outlaw droning” his clients’ properties, and I was amazed at the quality of the video. He took me along on one of his shoots and let me fly his DJI Phantom 2 Vision, which was the model just before the plus that did not have the 3-axis gimbal.

Though the images where sharp, the video got shaky during turns and when it was buffeted by wind gusts. Shortly after that, he upgraded to the DJI Phantom 2 Vision Plus which included the gimbal and a higher resolution camera, delivering full 1080p HD video. Once I viewed that stabilized aerial video, I was convinced that it was the platform for my market research.

The second DJI feature that sold me was the "return home," which enabled the safe return of the drone if the control signal was lost. I knew that was important, but I later learned that having an emergency retrieval capability was critical. Over the years, I have lost count of the number of times I have lost communications with my bird due to numerous factors, and it simply flew right back to me. Without that automatic return, I would not have invested the $1,300 in a flying camera.

2. DJI UAV Dynasty

DJI was started by Frank Wang in 2006 from his Hong Kong university dorm room. Over the years his perfectionism, reminiscent of Apple's Steve Jobs, launched DJI to become the world's largest UAV provider, holding 70 percent of the consumer drone market share by 2015.[1]

The DJI golden quadcopter line, the Phantom, started in 2013 with the Phantom 1. This was improved upon in less than a year with the Phantom 2 featuring a game changing capability called the Intelligent-Orientation-Control (IOC), enabling the UAV to practically fly itself. Later in 2014, the $1,299 Phantom 2 Vision Plus hit the market, touting the 3-axis gimbal and the integrated HD video camera filming at 1080p and capturing 14 megapixel (MP) stills.

For more professional aerial photographers the DJI Inspire 1 was made available with a 4K camera and more robust flight controls. It was a much larger quadcopter than the Phantom both physically and pricewise (prices started at $3,000).

The insides of the Phantom 2 were redesigned and deployed as the Phantom 3 hosting numerous upgrades, including a 4K camera, Lightbridge communications technology, active motor braking, optical-flow stabilization, and a new DJI pilot app with most of the Inspire 1's flight control capabilities.

Only since 2013, DJI has come light years forward in consumer grade UAV platforms delivering professional level aerial photography.

1 "Rise of China's Drones," Yu Dawei, Caixin, accessed November 2015. www.slate.com/articles/technology/caixin/2015/07/drones_in_china_can_the_country_s_industry_for_uavs_bloom.html

Combine that with their dominant market share and ever-increasing vendor participation, and DJI could be called the Microsoft of the UAV market. It is for this reason I am focusing the book and directing aspiring commercial drone pilots and business owners to invest their time and efforts to enter this exploding marketplace with a DJI product. No, I am not getting any financial benefits endorsing DJI. I just know what has worked for me on a consistent basis in the air delivering awesome HD video and high-resolution photos to me on the ground, with my bird coming back to me every time.

Product support is critical with this high technology, which involves complex hardware and multiple entities of software that all have to work flawlessly for flight safety and mission completion. When you really think about it, these DJI UAVs are flying computers with a primary job of aerial data acquisition. The data is video, digital images, and aerial telemetry that can be stored in onboard Micro SD media, simultaneously transmitted to ground-based platforms such as iPads and smartphones, or downloaded to host computers after landing via USB cable connections.

3. Hobby Shop Support

Making all these flying parts work on a consistent basis so you can make money monthly takes a good support organization. DJI has a strong network of authorized dealers and after-sales support vendors all over the world. As of mid-2015, there are nearly 20 storefront hobby shops in the US that can directly support you over the phone and repair your downed UAV at their store.

I have been working with UAV Direct out of Austin, Texas, for more than a year now, asking numerous questions and getting great advice. Nathan and Eric are top-notch drone techs and pilots offering training, sales, service, and parts for all DJI UAV platforms.

US-based drone support outfits like this are critical to keeping your birds in the air and in good shape. They can also help train you in safe UAV operations and teach you how to get those beautiful aerial shots your future clients will be expecting.

I have used online groups such as PhantomPilots.com for discussions focused on the DJI line of UAVs. I learned a ton of information browsing the site subject threads, which opened my eyes to the numerous capabilities my Phantom 2 Vision Plus had that I wasn't aware of. It was on this blog I learned about the other non-DJI endorsed vendors out there that repair and/or replace broken gimbal/camera assemblies, since that is the most crash-vulnerable component to the whole Phantom 2 configuration.

I know this because the first week I had my original Phantom 2 Vision Plus I was jacking around with a buddy, and I was in a rush to film him as he drove away down my rural 1,000-foot driveway. Not paying attention, I accidentally flew my brand new bird into a live oak tree, breaking the tender, small ribbon cable connecting the camera to the gimbal. I wanted to puke.

After I chilled out with a couple of beers, I went onto Amazon.com and purchased a new camera/gimbal set for $699, and swapped it out when it arrived a couple of days later. Since then I've found other vendors that will fix a broken camera/gimbal combo for far less.

It was a hard lesson learned, but those UAVs do exactly what you tell them to do. Be careful. Do not get rushed. Keep your head up and your eye on your bird. If you have UAV problems or need upgrades, the DJI support infrastructure is out there and ready to help you keep flying.

If you are going to get into UAVs, you cannot go wrong with a DJI Phantom 2 or 3 aerial platform. With such a large market share, proven track record in the air, spare part availability, and very good after-purchase support, you cannot go wrong starting your commercial aerial photography business with this line of UAVs.

4 Developing a Commercial UAV Business Plan

Before you purchase a drone for potential commercial use, take the time to lay out your strategy by developing a formal business plan for your new UAV-based company. Having a written plan is a must for pursuing any start-up venture and really helps you organize your thoughts, plans, and figure out the project finances.

Let's look at a generic outline of a business plan in Sample 1 and begin to insert the particulars for setting up a small aerial photography drone company. You can work on putting your plan together as you finish reading the rest of this book.

Take this basic outline of a plan and adapt it to your UAV company dream. The primary goal here is to get you to write out a plan. It will save you so much heartache in the long run and help you to plan for the route, the supplies, the end goal, and solutions to the problems you may encounter on that road. Flying UAVs is fun and we want to make money doing it. Therefore, we need a business plan to merge fun and money.

Sample 1
Business Plan

Executive Summary

The initial section lays out, in a single-page statement, the structure of your company, the services and/or products you will be marketing (aerial HD videos and high-resolution photos), how you will be marketing it (website, handouts, direct cold calling), capital requirements (where the drone purchasing money is coming from), and future growth plans (more drones? More pilots?). Although it comes first, it is usually written last as it summarizes the rest of the sections of the plan.

General Business Description

Write a more detailed section to articulate a mission statement and company objectives. Explain why you want to start the company and what you want it to accomplish over a specified time frame. If there is even a remote possibility of obtaining external financing of this new UAV company, this will be the guts of the plan that a banker or even your rich father-in-law would want to review. Be sure to specify the type of company it will be: An S-Corp, LLP, or LLC, etc.

Products and Services

List both the obvious deliverables, such as aerial videos and photos, that will be the default products of the proposed company, along with future services such as 2D and 3D mapping, agricultural temperature and moisture monitoring, and/or other aerial data acquisitioning operations. Create initial pricing data by assigning values to your airborne efforts. For example, write down how much you are going to charge for creating a production level HD video of a 15-acre property that is for sale.

Marketing Plan

This is probably the hardest part. You can have the coolest product or service ever, but if you can't market it correctly, you'll never earn a dime from it. Take plenty of time gathering data for your secondary research phase from every online source you can muster. Then go to physical locations, such as libraries and Chambers of Commerce, to sponge up information. Then it's up to you to conduct primary research by hitting the bricks and talking to people, including potential clients, other UAV pilots, drone blog members, and current small-business owner-operators. Put some real effort into this section of the plan with statistics, numbers, costs, and forecast data in the following marketing plan subsections:

Economics

You have to know your target market. That includes having a realistic expectation of the financial health in your area, along with the demand of that local market. Also, you will have to identify the trends in the market to be able to handle the barriers that you will encounter.

Your Products

You are offering aerial photography attained by UAVs and delivered in HD video and high-resolution images. You will have to list the specific deliverables that will be in your offerings. Beyond that, it will be critical to articulate the benefits and features of your products to illustrate how they will be superior to the competition's offerings.

Customers

Now that you have services to sell, you have to identify your potential customer base. This will be highly dependent on where you live and how far you are willing to travel. Group potential customers and industries based on the difficulty of entry. Mom-and-Pop shops are much easier to approach to get an initial gig compared to the complex entry process required to get on the approved vendor list of a Fortune 500 company.

Competition

Here is where the drone wars begin. Even though the UAV commercial market is just getting started, commercial droners are multiplying fast. Hopefully you are in a customer-rich and UAV-pilot-poor market so the competition level is low. Either way, you will need to know who you will be vying against for each commercial flight.

Sales Forecast

After compiling the previous data points, put together a conservative sales forecast as a financial goal. This will help you organize your efforts and hopefully be able to pay for each step of your success. Come up with a "best case" and "worst case," but shoot for the middle of those two roads, with your mandatory expenses and bills based on the "worst case" scenario.

Operational Plan

This section of your business plan should desribe how your aerial photography business will operate on a day-to-day basis. This is a plan on who does what.

Production

Describe your steps here to fly the sites, download and edit the media, and deliver the product, which will be comprised of HD video, high-resolution photos, and 2D/3D maps. Know what drone you will need, what critical accessories like batteries, and the delivery media for the client such as USB drives, Dropbox, and YouTube.

Legal Issues

Hopefully by the time you are reading this, the FAA has its stuff together so all you have to do is take an online test and pay $100 for your commercial UAV pilot's license. If not, get ready to do the 333 exemption dance like the thousands of early commercial drone pioneers did. Either way, go legal on the federal, state, and municipal levels so you will not have to look over your shoulder, and make sure you get the right kind of drone and business and any other insurance you need.

Personnel

Initially it'll just be you with your many hats to wear, and an observer as needed. However, if things really take off, you may need more pilots, sales personnel, and a bookkeeper. If your business plan is aggressive, then this is where you line up your headcount growth chart.

Inventory and Supplies

How many drones do you need? That may be the question you first hear from your wife, but have a good answer for it. In my gun collection, every firearm has a job; it should be the same for your UAV fleet. Each drone will have a range of capabilities and others will be backups. Remember, in the real world, two is one and one is none. Have spares, ample supplies, a massive battery array, chargers, and replacements for any critical items.

Start-up Capital

This is the do-or-die part of the business plan. Most new companies fail due to insufficient funding during the first few years. In other words, they run out of money before the money comes in the door.

One of the best ways to get started is with substantial seed money to get you (literally) off the ground. You can save some cash on your own from your regular day job over time or you can get a loan from a source that shares your dream. Whichever source you acquire the funding from, make sure it is good for you in the long run. Avoid heavy debt from negative sources. It's OK to start small and work your way into more UAVs via more money later. You do not have to get a $100,000 loan and launch a fleet the first month. You should aim to get into the black, financially, as soon as possible. Also, give yourself a healthy financial buffer for stuff that goes wrong, because it does, and it will.

Redefining Your Plan

A year into the UAV business you may find what works, what does not, what you like, and what you don't. Dealing with real estate agents may get old fast, but working with construction site supervisors may be a joy. Either way, odds are you will be tweaking your original business plan on the fly just a month into it, so be ready to do just that. Remember, this is your company. You are the boss. You make the rules. Money is not happiness, but it sure is freedom!

5 Aerial Photography Markets

Financial diversity is a good thing. It is akin to not having all your eggs in one basket. Identifying and hopefully doing business in multiple UAV markets is also a good thing. The following industries are in dire need of low-cost aerial photography and are the best sources of immediate income. The key to prosperous aerial photography is repeat business. It can be fun making money creating a really cool aerial HD video of your buddy's ranch out in the country for a couple hundred bucks once. It's quite another prospect to make a deal scheduling monthly, if not weekly flights, over an ever-changing environment such as construction projects.

Let's step through the top markets and identify their needs for our UAVs.

1. Real Estate

Real estate is the easiest industry to handle when you're dealing with properties more than five acres in size, and it can offer the most repeat work. Think about it: When you pull up to a suburban property for sale you see probably 90 percent of the front yard. Then go into the house, step out the back door to see the backyard, and you have seen pretty much the whole property. Residential plots with acreage need a more comprehensive presentation. This is especially true when we look at rural properties such as family farms, weekend rural retreats, and full-fledged ranches.

House and Acreage

Riding down the road pointing at fence lines, walking through trails, or taking a ride on a Polaris Ranger around the property line can be very enjoyable, but that takes time and effort. It is also difficult to grasp the overall coverage of the property by just driving or walking around. You can use an aerial tour to help grab the attention of the browsing potential buyer and to close the deal on the estate.

Do not waste your time offering your aerial services on urban or even small suburban lots of fewer than 10,000 square feet. Focus your marketing efforts on at least 5-acre properties that are for sale. The larger the better, since the need for aerial photography grows with every acre. It will be nearly mandatory soon for any rural property more than 50 acres for sale to have a production-level aerial HD video with scores of high-resolution pictures. The sweet spot will be rural properties from 10 to 100 acres, since they comprise the majority of farm and ranch parcels for sale.

There are numerous consumer-based real estate websites to find sales leads. These Internet gold mines are MLS-like websites such as our TXLS.com here in Texas. I choose the central Texas area on the search tool and pick all rural properties for sale that are 10 to 100 acres in size. From those hits, I make a spreadsheet of the locations and realtors selling those properties, which makes up my sales call list. As long as my price is competitive, the agent's response is usually positive.

2. Construction Projects

Construction General Contractors (GCs) are all about status reports and visual presentations to the clients for which they are building infrastructure. GCs have regular meetings with their customers in a continuous effort to keep them informed, advise them on progress, and explain delays. Having up-to-date views of substantial construction projects in a visually pleasing, aerial format can give them the "warm and fuzzies"; they can see that all is proceeding as planned.

Construction Site

Your production-level aerial HD videos also enable these GCs to have excellent marketing material for future bids and sale presentations. That is why the construction industry is an awesome market that is just now getting tapped.

One good way to find construction prospects is to drive around and look for vacant lots or demolition sites with large signs displaying that a contractor is about to break ground and start work. Timing is critical on projects like these, so jump on them and be the first drone in the air.

The potential for repeat, consistent aerial projects is tremendous in the construction market. Nail your first project with smooth videos showing the building process in HD video coupled with time- and date-stamped, high-resolution aerial photos of the construction stages over time, and you will likely be in demand for future projects.

3. Agricultural Services

Numerous drone articles have stated how the most lucrative UAV market will be the monitoring and managing of agriculture with drones as flying crop monitors. UAV technology is already extensively in use all over Japan, since they have limited land resources and have to be extremely efficient in all agricultural projects. At least 10,000 drones are in flight daily over Japanese farms armed with infrared cameras sampling temperature and moisture, and taking photographs to feed into sophisticated software. This enables farmers to have round-the-clock metrics on the health of their crops.

Markets like this are just emerging, hard to enter, and require substantial investment in both UAV platforms and software applications to handle all the complex data. Fixed-wing drones are more likely to be used in this market due to the massive size of the areas that need to be flown and mapped. However, smaller operations could be serviced by simple quadcopters like the Phantom.

Location can be a complicating factor if there are no crop-producing farms within an hour of your residence. My county primarily raises only cattle and hay, with little consumable crops such as corn, cotton, and barley are produced. But, I'm already making contacts in surrounding counties that are crop-producing environments so I will know what their future agricultural aerial needs will be. That way, I can prepare now for the investment in long-range UAVs, specialized cameras, and agribusiness software platforms to meet the demand.

4. Two-dimensional (2D) and Three-dimensional (3D) Aerial Mapping

Websites like maps.google.com and applications such as Google Earth are wonderful for viewing the vacation locations you plan to visit this coming summer. The downside of these aerial image sites is that the resolution is not great and the images are usually more than a year old.

Your Phantom UAV can change all that for your clients by providing a two-dimensional map with high resolution and today's content. For example, say your construction client has done months of work on a five-acre site and wants to show a high-resolution before-and-after image to a customer. You can fly a 90-degree straight shot down in a simple grid taking several dozen aerial photos, stitch them together as one large image, and present them an overhead that mimics Google Earth but is from that very day.

Other clients may need some sort of 2D map aerial measurements on a regular basis to help them inventory large areas of gravel, coal, or other masses of stock. There are some low-cost software solutions available for UAV companies, enabling you to measure large entities from the air.

Aerial Mapping

The next step in both complexity and profit is 3D mapping. The same grid flying pattern and even the same low-cost bird can be used, but the images are fed into a much more comprehensive software package to compute not only X and Y, but Z to calculate volume. So with 3D mapping capabilities, the drone company can not only provide the area estimates of a flat surface area, but can also provide the volume of those inventory piles by calculating the height or depth.

This 3D aerial mapping software gets expensive fast with rental rates at several hundred dollars per month. Expectations also increase due to the precision of those calculations being critical. Issues such as accurate ground reference points and absolute point cloud positioning (GPS exact within one centimeter) become mandatory. Another name for this aerial market is called photogrammetry and it is at the top of the UAV industry food chain.

5. Event Aerials

If you ever get a call to aerial photograph an event such as a cross country race or demolition derby, be forewarned that these UAV gigs are both high risk and high profile. Due to the possibility of a high density of people, presence of law enforcement, fixed time for the event, and other things moving around, numerous difficulties and challenges are introduced into the aerial photography project.

Event Photography

The golden rule is not to fly directly over people; this can be tricky when they are scattered around and your shot is near or around them. Your Landing Zones (LZs) can quickly get overrun and you may need to have crowd control helpers in addition to a spotter.

Make sure you understand the challenges of a large amount of onlookers, even in outdoor wedding events. It only takes one drunk guy to mess with you during a critical part of your flight, or some privacy rights activist claiming your drone has violated her just as you are landing.

The upside is you can make some serious one-time money and have the opportunity to film high-activity events, which can be edited into spectacular videos.

6. Vehicle Dealerships

Car, boat, and RV dealerships are businesses in need of our aerial expertise for two reasons. First, we can help them make some really neat commercials with our UAV HD videos and they will pay handsomely for just a five-second fly-back, rising up by the American flag over the new car inventory scene.

Car Dealership

Some of the bigger RV dealers have huge inventories; the large vehicles take up dozens of acres of parking lot. Our drones can be flown on a monthly basis simply to provide them an aerial inventory either in a video format or just a date-stamped high-resolution aerial photo.

7. Inspection Services

Think of all the cell phone towers you see driving down the highway on the way to work. Each one of those structures needs periodic inspections by some poor worker who has to climb and eyeball the towers for any abnormalities.

UAV-assisted inspection services are taking off because it nearly eliminates the tower-climbing risks. The same goes for water towers and span bridges since they need yearly, quarterly, or even monthly inspections. Make contact with local engineers, cell tower owners, and area utilities to see if they can use regularly scheduled aerial videos and photos of their infrastructure to find faults before climbers are deployed.

The next time you hear a hailstorm, it may sound like opportunity calling your name. Some insurance companies are seriously looking at drones to provide high-resolution photos to aid in roof inspections. The larger the roof, the bigger the need for aerial inspection services, made affordable by small UAV quadcopters.

Tower

8. Emergency Services

If you are willing to fly at a moment's notice, then check out the possibility of offering your UAV services to Law Enforcement (LE) agencies and the rescue units for local fire and emergency services. Examples of LE drone needs span from searching large plots of land for illegal pot growing operations, to accident documentation flights, to SWAT aerial surveillance missions. The ability to arrive on scene and launch your drone within an hour can net you serious billable flight hours.

During my market research pro bono flights, I had the opportunity to participate on a field scan for a reported marijuana grow field in central Texas. We had to covertly enter, pick a hidden landing zone (LZ), and then quietly take off. From there we scanned at 200- to 300-foot elevations to avoid shotgun range looking for tents, water hoses, and the actual plants. The main thing I learned on that mission was to have at least a one-mile range with high-definition First Person View (FPV) in real time. With our Phantom 2 Vision Plus that was just not enough. But with the new Phantom 3's 1.5-mile range and full HD real-time viewing on the iPad via Lightbridge Communications, we think we should be able to find some cartel grows on our next operation.

After a significant investment in an infrared (IR) camera, feeding to a FPV live display for emergency rescue professionals can make you the hero of the month if you are able to aid in the finding of a lost child from his or her heat signature at 100 feet. The FAA has been quite responsive to special permits for UAV searches for lost persons in emergency situations.

On the Way to Set up a Law Enforcement Drone Flight

Both LE and fire rescue units may discuss and even purchase UAVs, but it would take a capitalist to keep the drone ready to fly when that emergency call arrives. If you can make that commitment and close a sale for a fixed price fee, your monthly income would be supplemented and you would have some interesting aerial adventures.

9. UAV Market Summary

When discovering which markets you will target, start out with simple real estate projects and define your craft from there. After you have a handle on the whole process, from making the cold call to closing the sale and cashing the check from the successful UAV project completion, then utilize those newly acquired business and technical skills to conquer another market. Use each market as a building block to the next one to achieve income source diversity.

6 Initial Investments: What to Buy to Start Your Business

Before you buy your first drone, make sure you have put work into your business plan and have laid out a common-sense route to building your UAV business. This chapter can help you forecast your initial investment requirements for the business. It is not just a simple purchase of a single UAV, but rather the whole package you need, to be able to earn with drones.

1. The Phantom

Since we are focusing on the DJI Phantom series of UAVs, the recommendation here is to get at least a Phantom 2 Vision Plus if not a Phantom 3. Both have the integrated 3-axis gimbals and HD cameras, but the Phantom 3 can come with a higher resolution 4K camera, along with numerous other improved features if you get the Advance or Professional model. Your entry-level UAV should be at least the Phantom 2 Vision Plus, since you can be easily earning with minimal investment.

Try to avoid immediately attempting third-party gimbals and cameras like GoPro HEROs. These cameras can deliver incredible pictures, but getting all the multi-vendor components to work well together complicates the process. It is simpler to have a single solution with a fully integrated power system, HD camera and gimbal combo, and one vendor for simple software management and control systems, at least until you get going.

Phantom

Avoid buying your Phantom used from a buddy or from eBay. You really want a brand new drone to go commercial. I recommend purchasing it from Amazon.com and choosing the option for insurance for less than $100 like I did, just in case.

2. iPad Flight Controller

I quickly found it impractical to fly a Phantom with a smartphone due to screen-size restrictions. It was nearly impossible to see the ever-important power level in the upper-right corner of the DJI Vision app on the iPhone. Even the larger Android Galaxy 4 and iPhone 6 Plus are not large enough to see critical digital avionics and the clarity of the FPV images.

Next I tried larger Android tablets, but even they had a small font hard-coded in the DJI Vision app which made it hard to see important flight numbers. Only the full-sized Apple iPad gave me enough screen real estate to have a decent view of both the flight data and what the camera was displaying.

Though iPads are more costly than Android tablets, the iPad's battery life is outstanding along with DJI's solid port of their Vision software. Since there are really no significant moving parts in an iPad, you

iPad View

can safely buy it used as long as the battery is good and the screen is clean. All you should have to do is update the operating system and install the DJI Vision app. Make sure you purchase an iPad with cellular capabilities and the GPS feature, so you can utilize the dynamic home point return feature. This will enable the DJI Phantom to land wherever you are.

Another reason to go with an iPad rather than the iPhone is that you may have to use your iPhone for different functions while your iPad is running the DJI Vision app flying the Phantom. Yes, there is probably an FAA rule about flying and talking on a handheld device, so let's not fly and text!

3. Phantom Accessories

The first accessories you purchase should be extra batteries. One thing is for sure: You need numerous good, dependable, and genuine DJI batteries. I pack six Phantom 2 batteries in my travel case, and there have been numerous projects where I have used all of them during a single flying session just to shoot all the scenes I wanted correctly.

Another reason you need extra batteries is you may ruin one or more by poor charging practices. This part is critical; please remember what I'm saying here. If you charge your batteries on a Friday night for a Saturday flight session and some weather problem or scheduling change cancels that event, you will need to discharge the batteries rather than let them sit in the case fully charged for several days.

DJI batteries have a nasty little secret of "cooking out" if they are left in a 100 percent charged condition for ten days or more. One or more of the inner cells goes bad, and you will not know this until you are up in the air and have a rapid discharge. This problem can take down your bird's power from 50 percent charged to 30 percent to 20 percent and lower in less than two minutes! If you are at a critical flying session and you have a rapid discharge event, you will need to swap batteries so you can keep filming. Some batteries will last longer, so label them and keep good records on which ones have the best flight life.

Other accessories you will need include serial chargers to charge multiple batteries one after another, spare Micro SD cards, a DC power adapter for your car so you can recharge your Phantom gear, and a set of spare propellers.

4. Shaded Flight Deck

The first thing I looked for on the web was a holder for my iPad and Phantom remote controller, with a shade and a neck strap to hold it all so I had my hands free. I ended up making one out of plastic sign material, which has worked out just fine. You can build one yourself or look for something similar on the web that you can purchase. The point is to get something that will hold the iPad and controller in the shade so your hands are free to manipulate the UAV and take care of other tasks.

Shaded Flight Deck in Use

The shaded part of the flight deck is critical, since direct sunlight will not only make it hard to see the images and information on the iPad, but it will heat it up and shut it down in minutes when directly exposed.

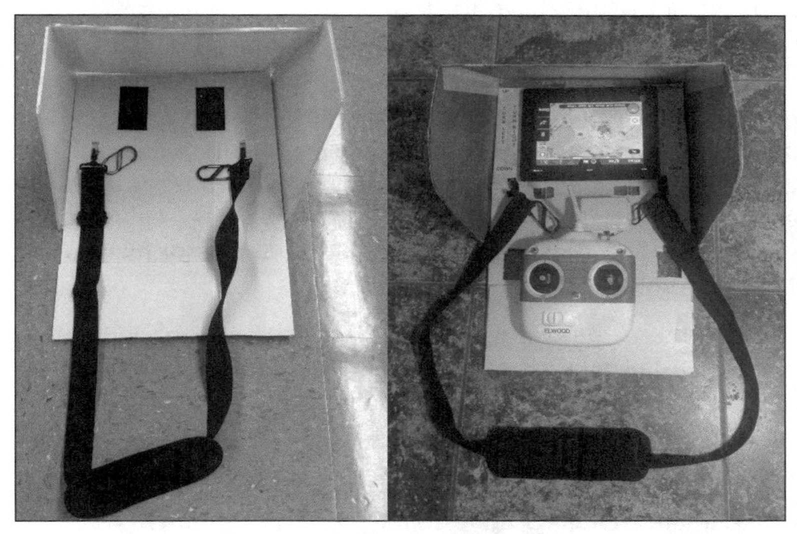

Homemade Shaded Flight Deck

Another use for my shaded flight deck is to paste my credentials and notifications directly on it so onlookers are alerted I'm a licensed commercial drone pilot and to stay clear of our LZs. The deck should be able to fold up and fit nicely in your drone case which will be discussed next.

5. Travel Case

Having a custom-fitted Phantom travel case is critical. You cannot be a commercial drone pilot without it. Purchase the hard case in aluminum, steel, or hard plastic, rather than the soft, backpack type of case.

The case is important, so you have a single container to house all of your necessary equipment to perform a commercial UAV project for the day. It should comfortably hold the UAV, controller, props, at least five Phantom batteries, USB cables, chargers, and your shaded flight deck, folded up. These hard cases will also serve as a flight takeoff deck in most circumstances, so you can avoid dirt, grass, and dew.

Having all your UAV gear in a single case is not only convenient but also practical. You want to pack up your drone case at night and place it by the door to be ready for the morning takeoff. Without the all-in-one drone project container it is too easy to forget critical components. Plan on spending at least $100, if not around $200, and make sure the gear all fits securely with nothing bouncing around inside the case.

6. Media Editing Platform

If possible, obtain a Windows PC strictly configured to edit and store HD videos and high-resolution photos from your drone business. Try to dedicate the capital for a computer with a solid state operating system drive, at least a two terabyte (2TB) hard drive for data (video and photos), 8GB of RAM, an Intel I5 CPU, Windows 7 Professional or Windows 10 operating system, and a large 24-inch or bigger LED display. Try not to use this PC to surf the Internet, use it for other work or fun projects, or let the kids do their homework on it. It needs to stay pure and run fast.

In addition, you will want an external USB 2TB backup hard disk drive (HDD) for data backups, a battery backup (UPS), and dozens of 4GB USB thumb drives for deliverable media to clients.

Plan on dropping $1,200 to $1,500 for strong workstation hardware.

7. Aerial Editing and Other Software

After you get the hardware, it will be time to obtain the software to really be productive. First you will need the latest Microsoft Office suite for proposals, spreadsheets, and other documents so set aside $250 for that. Next is the video editing software, PowerDirector 13, which you can get from CyberLink for $75 to put those awesome aerial videos together. The DeFishr program is a must to remove the fisheye effect from the videos of Phantom 2 Vision Plus cameras, but the 2D mapping stitching program called ICE from Microsoft is free!

Finally, background music for your videos is usually a must, and make sure it is copyright free. That is so you can put it on YouTube.com and other Internet sites without having to worry about copyright violations. I get original tunes with all rights from Pond5.com starting at $15 each. You only need five to ten scores in different moods to get your background musical library started.

This seems like a lot of extra stuff but it is all critical for planning, deploying, filming, editing, and delivering UAV-based aerial photography to your clients.

7 Weather Issues: The Good, the Bad, and the Ugly

The most difficult part of being a commercial drone pilot is finding paying jobs. The second most difficult part is getting the right weather to fly. I have had more days of flying rescheduled due to weather problems than days that went as planned. I spend many mornings scanning weather websites, viewing my favorite iPhone forecast apps, and watching the flag blowing at the end of my driveway.

To make money, you want to fly a project well and, preferably fly it only once. You also want to have smooth air and light that enhances the aerial view. The weather is your master. The only thing you can do is learn to plan, interpret forecasts, be able to move fast during a clear sky opportunity, and most importantly, have patience.

Too many times I got rushed, did not heed the wind warnings, and went ahead with my planned flight. Usually, this led me to bringing the drone down immediately due to gusty winds buffeting my Phantom, which overburdened the gimbal. The result would have been a shaky video. The projects where I waited for favorable weather conditions and preferred lighting worked out better, and I did not waste time or put my UAV in peril.

1. Weather Issues When the Client Has Time Expectations

You will have times when a client sees your past aerials, becomes very excited to get his or her ranch photographed, and wants it done immediately. This is where you have pull on the reins and set some time expectations. Explain to the client the difficulties with weather, which include wind, lighting, possibilities of rain, and even low clouds.

Let the client know that it could be several days or maybe even a couple of weeks before you are able to fly under favorable conditions. It can get even more complicated when only specific time windows during the day are available. You could have had a smooth air morning with good light at 9:00 a.m., but the client wants a sunset. That can be troublesome in the summertime, when afternoon thunderstorms and warm thermals are probable in Texas; always keep seasons and weather in mind wherever you are.

Other conflicts involving weather could occur when the client wants to be present for the aerial photography flight, because he or she wants to watch the flight and help direct what is filmed. The windows of availability, coupled with good flying weather, can be very difficult to synchronize.

Work with the client and explain how you need flexibility to adapt to ever-changing weather conditions. Some areas of the US are more weather-friendly than others. Flights planned in sunny, southern, inland California can be much more predictable than those north of foggy San Francisco. The same goes for times of the year; for example, here in Texas when we get a high-pressure cell parked over our state for weeks, clearing our skies and calming our winds, it is time to fly!

2. Smartphone or Tablet Weather Apps

Mobile weather applications are critical for the commercial drone pilot. We have to have the most up-to-date forecasts by multiple sources, so we can take the average of their educated guesses and hope to be right. As weather can change on a dime, our need to stay informed on atmospheric developments during UAV flight windows is very important.

Fortunately, there are numerous smartphone and/or tablet weather apps available to meet those needs. This is especially true for iPhones and iPads, with apps from the Apple App Store. Some are free, some are inexpensive. My favorite weather app is "The Weather Channel Max," which has the icon of TWC Max. For planning, I use the 10 Day forecast to target a good sunny day with minimal winds. On flight day, the first thing I use is the Hourly tab to pick the time to fly or see if my fixed time window is clear.

The Weather Channel Max App

For critical storm-dodging information, I use the iPhone app RadarScope. I used that jewel during my tornado chasing days back in 2010. This $10-a-year subscription level app did an awesome job helping me find funnels in Oklahoma, and now I use it to help me see the bad weather coming. It will display storm tracks and lightning strikes headed your way so you can avoid them.

Another handy weather app to have in your arsenal is RainAware. This GPS-based app will tell you to the minute when rain from a storm will arrive at your location. I have cut it close on a few flights trying to get that last sortie in just to make sure I got the video I needed. Once, this app displayed that I had 30 minutes until a 70 percent chance of rain hit. I used the window, got another fresh battery in my Phantom, got the bird up, got the shot, landed safely, packed up the case, and it started raining within 3 minutes of RainAware's calculation.

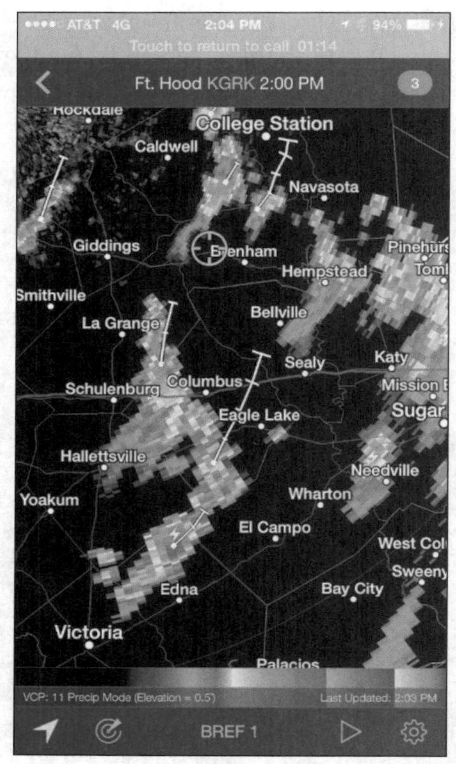

Radarscope App

3. Avoid Winds and Rain

Not flying in the rain seems obvious, but when your UAV is more than half a mile away over a pasture, and a fast-moving storm develops out of nowhere behind you, avoiding rain can become an emergency. Knowing the predicted chance of rain for that area during that time window is crucial for planning and must govern your plans.

For planning purposes, don't plan on flying if there is more than a 50 percent chance of rain. Watch the forecast carefully, with optimism, if it is 20 to 40 percent, and plan for a strong probability you are flying if the forecasted rain is 10 percent or less. Utilize all three of the previously mentioned apps to closely monitor the conditions when your bird is in flight. Remember, just a few raindrops inside an electric engine can take down your UAV, and if the internal circuits get wet, the whole bird can fry.

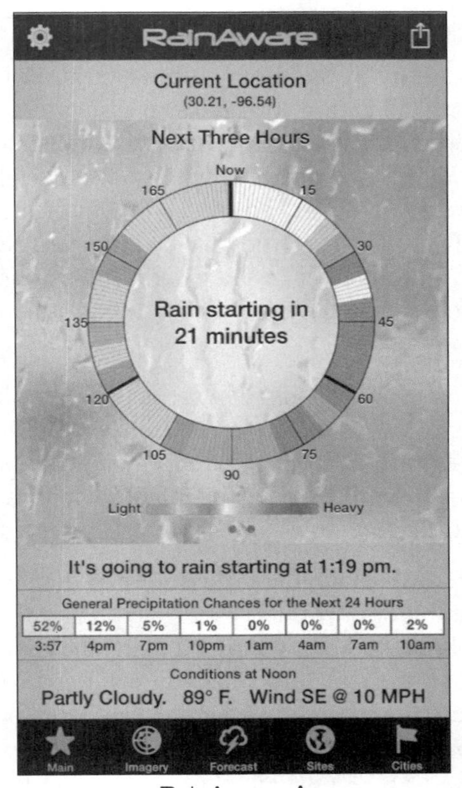

RainAware App

Though wind will not likely destroy your drone as fast as rain can, it can spell trouble for flight stability, and high winds can endanger it. My rule of thumb is to avoid average winds in the teens. I'll fly in 10 to12 mph winds, but will opt out in the 13 to15 mph range. Any wind higher than 15 mph is the line for grounding for my UAV fleet. Though Phantoms can fly in winds that are 20 to 25 mph, you are risking the UAV and pushing safety thresholds.

High-teen winds push the engines, strain the props, drain the batteries faster, and make for poor aerial video. The 3-axis gimbal can handle winds 10 mph and less just fine. However, less wind is always best. When my driveway flag is hanging down perfectly still, I want to have my bird in the air earning.

4. Seasonal Climate Aerial Factors

It is not just the day-to-day and hour-to-hour weather limitations that should be monitored, but the longer term issues regarding seasons need to be considered. In Texas, there are two times of the year when the ground looks brown and dead, which is not especially photogenic. During what seems to be our seven weeks of winter from mid-January until the end of February, the trees have no leaves and the grass is weedy brown. Six months later in the peak of our five-month summer with no rain, the trees may have leaves but the ground is again brown and cracked.

In the springtime, most everywhere in the US will be your best business flying season due to moderate temperatures and blooming foliage. The March-April window is bluebonnet season here in central Texas, which makes UAV pilots clear their calendars for clients requesting springtime flyovers.

The beauty of leaves changing their colors in the fall can be another aerial photography opportunity, especially in the northeast before the snow starts falling. To keep your bird flying and earning on a consistent basis through the calendar year, you will need to diversify your client base. Strive to serve multiple industries with busy times that have little overlap. In other words, make sure you have enough aerial photography work from various clients to avoid the feast and famine pitfalls into which many small businesses fall.

8 Drone Operation: Learn to Fly before Taking Photos

Always remember that you are a pilot first and a photographer second. Your primary responsibility as a commercial UAV pilot is to safely operate and land the your drone without incident. The key word here is "incident," since that is what triggers an FAA investigation which, best case, results in a major loss of your time and money. If there is property damage, major negative press exposure, or injuries, your penance could be substantial and possibly life changing.

The point I am making is that you must be consistent in your ground processes, safety procedures, and aerial discipline to conduct every sortie (for the uninitiated, that's military aviator lingo for take-out-mission-landing), in a professional manner without any close calls, risky maneuvers, or any negative appearance to the general public. The UAV industry has been hammered enough already by those who irresponsibly fly their drones into fragile or restricted environments, generating bad press for fellow pilots.

1. Configure Your Shaded Flight Deck: Prep Work

To get started flying your commercial Phantom right out of the box you need to pause and do some prep work first. As I mentioned in the initial investments chapter, the shaded flight deck is a must. Since nearly all your flying sessions will be during the day with sunlight reflecting off your iPad, shade is critical. Keeping your hands free to manipulate the UAV is important along with being able to utilize your phone for other technical chores tied to the aerial project. All these free-hand requirements are made possible by the shaded flight deck hung from your neck.

You can use the following Homemade Shade Flight Deck Instructions to construct your own shaded flight deck before even beginning your flight training. This may seem like a silly homemade contraption, but it will be a necessity out in the field when you are flying hours per day. (See Chapter 6 for more information.)

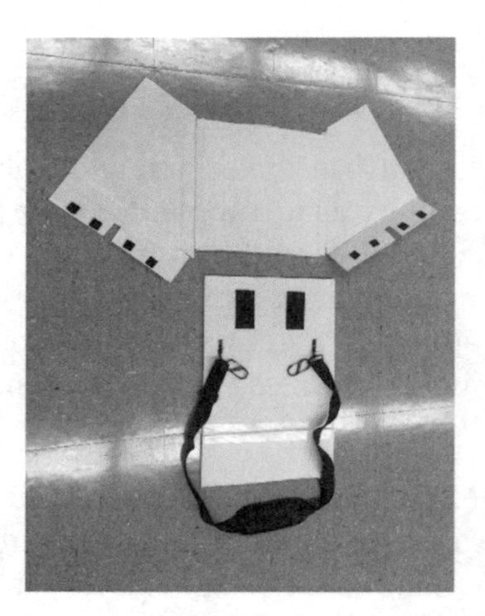

One of the major deck construction points is for it to securely hold the remote controller and your iPad. Utilize sticky-backed Velcro to hold the iPad to the deck, and mini-S carabineers to secure the Phantom's remote controller. Use an old shoulder strap from a laptop case to hold the deck hung from your neck and the lowest edge against your stomach. The shade should be removable with small-Velcro fasteners to connect and disconnect for quick set up and disassembly.

Homemade Shade Flight Deck Instructions

2. Preparing for Takeoff

Since this book is not a beginners' guide to flying drones, I will assume you have already read the Phantom manual and know the preparations, safety procedures, takeoff processes, and so on. What we are going to focus on is some processes to assure the safety of your drone and the efficiency of the mission.

I never take off from the bare ground. There is too much dirt, loose grass, moisture, and other stuff that may hit my UAV on takeoff. It is for this reason I always use my drone case as a launching pad for my Phantoms. Since the prop wash stirs such a high volume of muck and other junk that can dirty up the lens, compromise the gimbal, or nick the propellers, I want to elevate the bird above the crud and give it a clean and level environment from which to lift off.

Some aerial project LZs will mandate such take-off techniques since the environment could be nothing but two-foot-high grass. There have been other times where I placed my bird on top of the roof of my truck just so it has a better chance to not hit bushes and trees on the way up and out.

3. Landing

Landing a quadcopter, especially the Phantom series, can be difficult because of the narrow landing rails. Tip-overs resulting in prop hits are easy due to wind gusts and sloppy piloting. Any violent motion of the UAV risks the camera, gimbal, props, and similar fragile components.

I have already strongly recommended taking off from a manmade platform above the dirt, sand, and grass to protect the camera and moving parts. Landing a quadcopter can be just as harrowing as take-off. Due to the dangerous, high-velocity spinning propellers on a quad-copter, I cannot recommend catching the Phantom in the air as a method of landing. For liability's sake, I recommend that you do not. It is much safer to land it on a hard, flat, clean, and dry surface such as a driveway, parking lot, or large piece of plywood.

Let me be clear, I do *not* recommend landing the Phantom by catching the landing gear and shutting down the engines to prevent the bird from getting dirty or experiencing a turnover while landing on the bare ground.

3.1 Emergency landings

Just like in a real plane, the pilot in command (you) needs to always be scouting for an emergency LZ in case of rapid power loss, bird hit, prop loss, or engine failure. This is one of the reasons to keep the UAV in line of sight; you can quickly land it away from people, water, trees, and roadways.

Before I arrive on the site of the flight, I map out possible alternate and emergency LZs with Google Earth the night before. Upon arrival at the site, I will eyeball those pre-mapped LZs to make sure they are available and drone-landing-friendly, without risk to humans and with low risk to the Phantom.

If a low-flying airplane or helicopter enters your flight area, descend immediately as long as it will help avoid a collision. When you see a hawk or other large bird getting aggressive with your UAV, fly the opposite way and lower your altitude. Keep an eye out for fast-developing rainstorms, or cold fronts with gusty winds sneaking up on you and the airborne UAV. Be ready to quickly get your drone out of the air and on safe, clean ground when risk appears.

In the event of rapid power loss, fly directly to your nearest pre-mapped emergency LZs. Focus on heading rather than trying to lower the bird since that will happen soon enough. You may establish enough forward momentum to make it to a non-disastrous area that is not inhabited, wet, or heavily treed.

4. Learn Smooth Aerial Maneuvers

Once you are aloft, get used to flying in the aerial photography altitude range of 50 to 200 feet Above Ground Level (AGL). The minimum altitude will be at least 25 feet above the tallest structure in the area. If the highest things in my flight area are 50-foot trees, I am going to keep my Phantom at least 75 feet high. The same minimum vertical separation goes for buildings, power lines, and all types of towers. (This is the current rule for 333 exemption pilots.)

After you have established a safe minimum altitude, you can start flying the bird around and get used to the sensitivity of the controls. The key to good photography is small and smooth flight manipulations, so get into the consistent stick control with that in mind. Try to avoid harsh turns, abrupt stops, and quick changes in altitude.

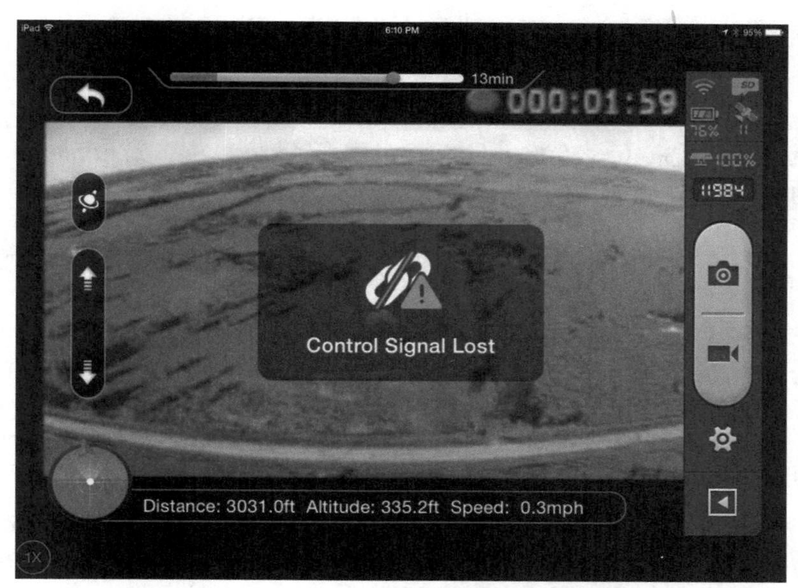

Control Signal Lost

Become familiar with what the bird looks like 200 feet away at 100 feet high then how it nearly disappears at 500 feet away and 300 feet high. Understand the radar simulator in the lower left corner of the DJI Vision app and attempt to fly it away from you, towards you, and then parallel to your position.

Discover how that battery consumption varies with how hard you work the UAV, and how fast 10 to12 minutes goes when you are flying around. See if you can learn to smoothly fly in a rotating pattern around your position, keeping your image as the center of the FPV frame while recording. Then train on doing clean, stationary rotations to get a 180-degree scan of the area. Practice going up in a slightly accelerated mode and then in a consistent climb mode.

Try all these different maneuvers over and over until you have confidence in your piloting skills. This may take 10 to 20 or more sorties over several days, depending on how many batteries you have and the time it takes to recharge them between flights.

The goal is to get your processes down to where you can configure your bird, get it cleanly up in the air, perform the aerial mission you need to gather the photography required, then return safely to the ground with no incidents. Practice is the only way to acquire this skill set.

5. Test "Return to Home" via Distance and RC Failure

The scariest test you will do when you first fly your Phantom will be testing the "Return Home" feature. This step is highly recommended since you will be using it either by choice, loss of communications event, or by controller failure. Either way, you need to test that it works and see how it works, and have confidence it will work when you need it.

Before getting started, make sure you are doing the following tests on full batteries, you have calibrated the compass (see your manual for that process), and the "home point" is correctly set to your current position. Then fly the Phantom about 100 feet up and 500 feet away with no obstructions around, preferably in an empty field.

Test one will be to simply turn off the Phantom's remote controller. Within ten seconds you will see a communications loss banner on your iPad DJI Vision app and the "Returning Home" notification.

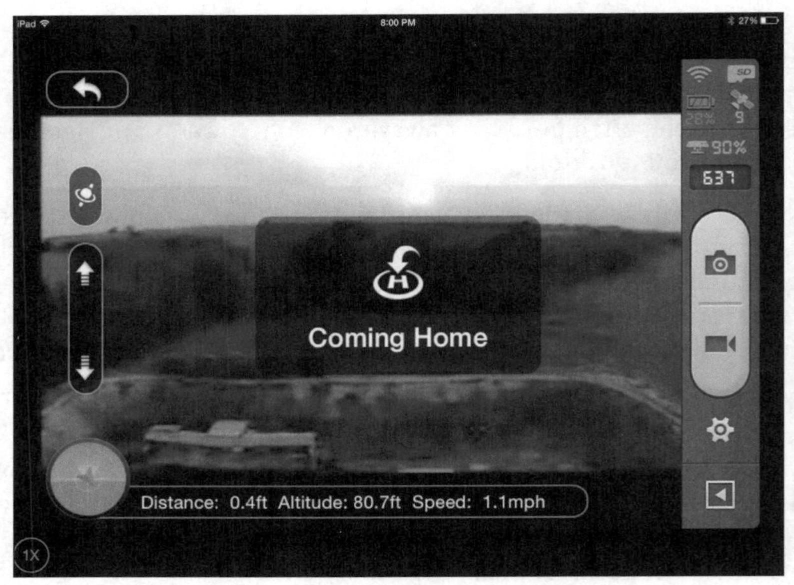

Coming Home

The avionic display should show the bird flying back to you and it should come back to your home point and begin its landing procedure. After powering the controller back up, you can toggle the top right switch down and back up which will return control if you want to take over the flight and abort the landing.

Test two (again, on a full battery) will be to send the Phantom up to at least 200 feet and start flying forward over a safe area. As long as you have good line of sight, keep going until the distance gets so far the communications link drops and the "Return Home" process starts. Note how far it got just before it lost signal. I have been able to get my Phantom 2 Vision Plus to just over 3,000 feet horizontal distance before it lost communications and returned home.

Verifying that the "Return Home" function will work in these controlled conditions is important so you can have the confidence in knowing your UAV will return from any treacherous flying environments during your real-world aerial photography missions.

These more advanced procedures and flying skills will be necessary to utilize your UAV as a flying camera in a professional and safe manner.

9 Flight Safety: Preventative and Preflight Measures

By default, you should already have RTFM (Read The Freaking Manual) and know all the safety procedures spelled out by DJI. We are going to cover some critical safety points anyway. As a commercial drone pilot you will be flying in many different environments, various places, odd times, under time constraints, with hundreds of people watching, or sometimes only just you and your observer for miles around. (More about observers in Chapter 11.)

Wherever you are flying, under whatever conditions, you are responsible for everything that happens when your drone is the in air. That might sound severe and overreaching, but in today's "fear-the-drone" atmosphere, combined with the few bad eggs flying UAVs into bad situations and the FAA clamping down on hobbyists while eyeing professionals, we all need to be extremely careful.

1. Know Your Environment before Taking Off

Besides weather issues as covered in Chapter 7, there are other things to know. Google Earth is your friend: It is free, and runs on PCs, Macs, and is available as an app on your tablet or smartphone. When you get a new gig in a strange part of town, the county, or even the state, check it out with Google Earth. Even if the image date is stale and more than a year old, it will help you plan the flight. Look for high power lines, cell phone towers, and other obstructions.

Make sure you are not near airports, by using the free app called Hover or the website www.mapbox.com/drone/no-fly. The huge five-mile no-fly radius the FAA has put on most airports can blanket some medium-sized cities when airports are within the city limits. Smaller, rural airfields without towers still may have a three-mile radius ban, but it is up to you to know about these restricted areas so research before you fly.

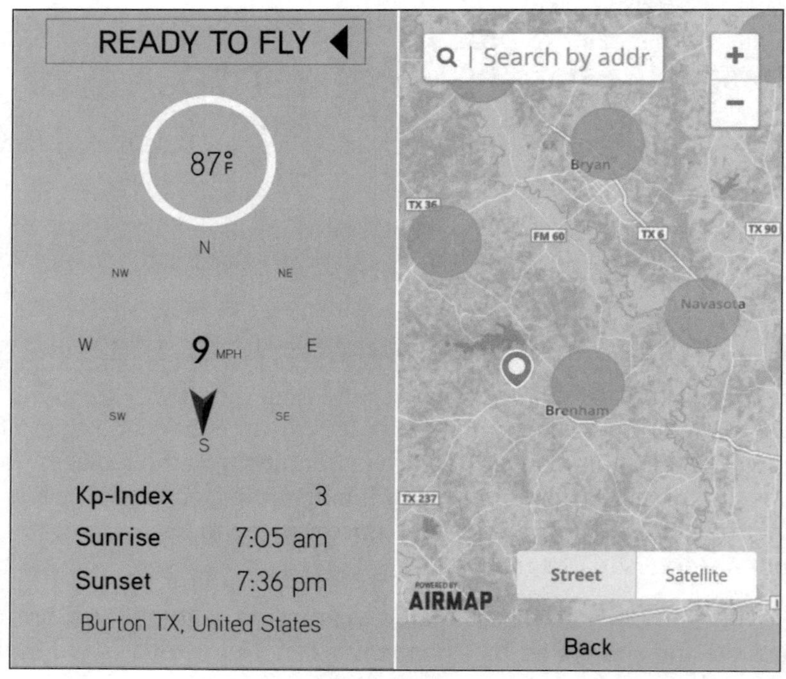

Hover App

There could be other restrictions for your target area, such as active forest fires resulting in wide-area UAV groundings. There could be city ordinances banning drone flights implicitly, which becomes a show stopper for your UAV project.

1.1 Site security

Check the crime statistics for the area. I have had to fly on the bad side of Houston a few times, and my observer had to double as my bodyguard. Good sources for identifying high-crime areas are www.CrimeMapping.com and www.MyLocalCrime.com, so use them. Being a concealed handgun licensee since 1996, I am always armed during my flights, just in case.

Situational awareness is mandatory not only for your airborne UAV, but what people on the ground are doing around you. Your observer will be doing double duty in high physical risk areas by keeping an eye on the bird, the gear at the LZ, and your back while the UAV is airborne.

Drones in the air can attract some negative attention. There have been cases where legal drone operations were being conducted and some buzzed dad of a daughter out by the pool thinks the UAV is aerial-gawking at his baby girl; trouble can begin real fast. Do not let hot-heads get anywhere near the drone. Keep it out of throwing range, and even shotgun range if necessary.

If bad guys come calling, you need to get your bird down quickly and safely, pack up the case and hightail it out of there. Avoid damage to your UAV or even downright theft of it by some punk who thinks he can just run by and swipe it out of the air or off the case as it lands. Come back another time to finish the flight with a more watchful eye (and perhaps a better criminal profiling ability).

2. Shoot-down Avoidance

Over the past couple of years, several nonmilitary, civilian-owned drones have been shot out of the air. Most of the time it was due to the pilot having to fly over personal property to get to another destination. The overflown property owner sees the UAV and takes its mere presence as a violation of his or her privacy rights.

Due to previously reported stories of bad drone operators purposely maneuvering UAVs to get aerial photos of girls by the pool and videos through home windows of people changing clothes, many citizens are quite wary of these flying cameras and think them nothing but an invading nuisance. This has grown into a problem where enraged property owners take vigilante action and resort to dangerous deeds such as using firearms to shoot down drones.

This action is against federal law, which forbids any nonmilitary person from shooting down ANY manmade aerial vehicle, whether it is manned or not; this covers all UAVs and drones no matter the size, shape, weight, or type. This means you as the pilot have a right for your drone to not be shot at, but you also have the responsibility not to put your bird in a situation that may brew up a problem.

As is said, "an ounce of prevention is worth a pound of cure," so keep your bird around 200 feet high when you are over potentially hostile territory, especially in rural areas during dove hunting season when nearly every country Bob has his 20-gauge shotgun close by. Do not make it tempting for Jimbo to take out a low-flying drone over his 15 acres, just over the next hill south of your land.

If you are performing commercial UAV flights for LE (Law Enforcement) and you need to fly in hot zones such as hostage standoffs, SWAT situations, and pot field aerial searches, then you are rolling the dice. As long as you are above 300 feet, you should be out of shotgun range since even 00 buckshot will lose its momentum after that distance. Handguns will be difficult to have the precision to hit at that altitude, and rifles such as AR-15s and AK-47s would be stressed to strike a hit on your UAV if you are moving horizontally as well as maintaining high altitude. Even if you are limited to 200 feet, sometimes UAV survival and the safety of people on the ground is worth bending the rules if illegal gunfire is involved.

Keep your ears open for anything that even sounds like a gunshot. At the first hint of gunfire, move quickly away from the shot origin. Then go to maximum altitude, which is probably hard coded in as 400 feet, and haul tail back to the safe LZ. If you are on a drug surveillance flight and you hear a shot, you and your LE buddy have been found. Time to get the guns ready, the bird down, and all of you out of the area very fast. Don't mess around with cartels.

Also on the subject of LE aerial drug operations with your UAV, remove all identifiers prior to flight to protect your identity for obvious reasons. Yes, you may be bending FAA 333 exemption registration rules (see Chapter 11 for more on the 333 exemption), but who wants the Zeta Cartel tracking you down from the November number on your Phantom?

3. Simple Onlookers

What we do is a novelty for now, and we have to deal with curiosity seekers on a regular basis. Nearly every flight I do in public I will see a person, usually a guy, come up to me while I have my bird in the air and start talking to me about the UAV. Sure, you want to be nice and share similar interests by discussing the model and its capabilities, but you are flying right now and cannot be distracted.

Things can happen quickly when your UAV is soaring around, and bad things happen even faster. You need to keep a hawk's eye on the drone itself, along with the avionic stats such as battery levels, altitude, and velocity, rather than yapping with a fellow geek about the industry.

Politely let the person know you really need to focus on the drone for the next few minutes and you will be glad to discuss UAVs after your bird is safely on the ground. Make sure the onlooker is not blocking your view to anything that could affect your line of sight with the drone and the LZ.

Observers come in handy here; have them intercept gawkers before they make disruptive contact when you are trying to fly.

4. Aerial Dangers

We have covered weather-related threats in Chapter 7, but it is worth repeating for you to be vigilant in atmospheric conditional awareness. Watch out for lightning, since it is one thing to lose a UAV but worse to lose a pilot. Flash floods can also be a danger to human life, even after you have secured the bird in its case.

Other aerial hazards include birds — real birds with feathers and claws. Some breeds of fowl will attack your drone out of territorialism, fear, or just plain curiosity. No matter what the attraction the bird has for your UAV, it can spell disaster if it hits a propeller and takes down your drone. I have had hawks, larks, and mockingbirds take after my Phantom numerous times.

During a trial long-distance flight over my ranch, some large bird attacked my UAV and the propeller killed the bird without the least flight disturbance. I saw a "poof" of feathers, the bird fell to the ground, and my Phantom flew home without a problem other than some blood on the propeller. I drove my Polaris around the drop site for a while to find and try to help the bird but I never found it. I'm telling this story because it could have come out worse in many ways. The bird could have taken down my UAV and I would have lost more than $1,000. Another scenario is the mystery bird that disappeared could have been a federally protected bird of prey such as an eagle, a hawk, or an owl, and I may have been subjected to a $10,000 fine from the wildlife feds. For the record, I believe it was a crow that hit my UAV. The point is, you as the pilot in command need to watch out for fowl that could either wreck your drone or cost you a fortune in federal fines.

5. Rapid Power Loss

We touched on rapid power loss earlier, but this is one of the most important topics in this book. Again, if you leave your primary batteries fully charged for over a week without using them, they will swell and seriously internally degrade. So much so, that the next time you fly with the compromised battery it may go from 50 percent charged to 30 percent to less than 20 percent in less than two minutes. Being over people, water, and trees would be a very bad thing if you lost power completely.

Prevention is the key to avoiding rapid power loss by draining out fully charged batteries within two to three days after charging them to 100 percent. For example, if you charge up on Friday thinking you are flying Saturday and plans fall through, do not let the batteries sit fully charged until next weekend. Take them out and discharge by doing a practice flight and get all the charged batteries down below 50 percent.

Recently I had a major gig and had six Phantom batteries charged up for a Monday morning flight, and it started raining Sunday night — for three days! Finally, I caught a break in the rain Tuesday afternoon so I did six sorties at my ranch lasting ten minutes each. The total time I spent de-charging all six batteries was over an hour and a half with takeoffs, landings, battery swaps, and power-ups.

To keep from going through this discharge procedure, keep your batteries de-charged and remember to charge them up no earlier than 24 to 48 hours before a flight. Sometimes I will only charge two out of my six batteries if I know for sure that it will be a short flight the next day. That way I will not have to empty out the four batteries I did not need. This preventative process is a royal pain in the rear, but the alternative is risking losing your bird or worse, it falling out of the sky and hitting a person.

6. Acquire UAV Liability Insurance

Stuff happens. Even when we take every precaution, have the proper licensing, take the required training, and have invested in top-shelf UAV equipment, things can go wrong. Since this is a business, you need to have liability insurance for your commercial drone's possible bad outcomes. Whether from rapid power loss, a snapped prop, an engine seize-up, a software glitch, a freak gust of wind, or just a simple human error, UAVs do come down in a negative way sometimes.

Murphy's Law is always in play, and if your UAV drops and damages another's property or injures a living being, there will be a price to pay. As a real, commercial drone pilot with either the early FAA 333 exemptions or the hopefully upcoming drone licenses, liability insurance will be a standard requirement for most clients to contract you to perform aerial services with your UAV.

It is common sense also. We have liability insurance on our ground vehicles. Private pilots have been dealing with airplane insurance for more than 100 years. The same should go for drones since they can fall out of the sky resulting in property damage and serious injury to persons.

You can rest assured that slimy personal injury lawyers are salivating to be the first attorney to sue a commercial drone operator. Protect yourself by first setting up a corporation or LLC as we will discuss in Chapter 20, but plan on getting liability insurance for your UAV operation before you ever take a dime for your air time.

Here are a couple of UAV insurance companies that may underwrite your drone business:

- Costello: www.aviationi.com/DroneQuotes.htm
- Drone Insurance: www.drone-insurance.com

7. Preparation and Preflight Checklists

Most of the technical problems I have experienced in the field could have been prevented with better preflight preparations the night before. Sometimes the smallest detail can have a domino effect, creating show-stopping difficulties if your flight planning is poor.

7.1 Night-before procedures

It may sound obvious, but remember to charge all required battery systems. These not only include the ample number of primary Phantom removal batteries, but also the rechargeable Wi-Fi range extender, remote controller, and the iPad. My rule of thumb for charging primary batteries is to charge one for every ten minutes of estimated flight time plus a backup. With that formula for a medium project requiring three sorties each running 10 to12 minutes, I would charge four batteries. On full weeks where I have daily flights, I would charge all my primary batteries and sequence through them during the following couple of days. That way I have ample charged cells ready to go, but none set for too many days risking the cook and swell problem.

I purchased a serial charger for the Phantom's batteries that would charge four over a two-hour period, one at a time. With that style I did not have to remember to swap them on and off the charger three to five times. It would be great to find a charger that can simultaneously charge four or more batteries at once, but I have yet to find one.

The older Phantom 2 remote controllers lacked the internal rechargeable battery and instead had four replaceable AA batteries. I keep the high-end lithium batteries in the controller with another spare set of four still in the package, stored in my flight case.

Before each flight I top off the extended Wi-Fi unit and the iPad, even though it usually takes two to three flights to get below 50 percent power. Just to be on the safe side, I keep a 12V DC to 120 AC converter in the car with extra charging cables. Those handy accessories enable me to recharge batteries on the road, in between flight sites.

Make sure you have enough space on your Micro SD card for the next day's flight. A good practice is to always download your videos and images to your dedicated video editing workstation immediately after returning home from a flight. Since your workstation should also be configured to perform backups that night, you can delete the last mission files from the Micro SD card, keeping the UAV's storage free and ready for the next photography project.

7.2 Secure flight case items

After all batteries are charged, the night before the flight, I pack everything I need into the flight case and verify the camera/gimbal lock is in place. That little plastic lock is important to put into place every time the drone is shut down, since it keeps the camera from swinging around on the loose gimbal. If anything happens to the tiny vulnerable ribbon cable, the camera system is shot.

The lens cap that comes with the Phantom 2 Vision Plus is very tight fitting, and I have nearly broken the gimbal by taking the cap off far too many times. I felt the risk of using it and possibly damaging the assembly was greater than the protection it provided, so I stopped using the cap. Since I constantly keep my Phantoms in cases when not in use, I have had no problems with dirty lenses.

Tie down your loose spare props, batteries, and charging cables so nothing moves around during your travels. When the case is shut and you move it around, there should be no sounds of items shifting around inside. Use extra foam, rubber bands, or Velcro strips to lock down stuff inside your case.

Place the flight case by the door that night (or in your locked vehicle that is parked in your locked garage), so the next morning all you have to do is check the weather and grab it as you head out to the flight site. If you have a sedan, use the trunk for security and for a truck or SUV, use a blanket to cover the case. It would really be unfortunate for some car burglar to break into your vehicle and get your complete aerial infrastructure.

8. Onsite Setup

Upon arrival at the flight site, get familiar with your surroundings and compare them to what you prepped while viewing Google Earth the night before, to find anything new that may risk your UAV or change the flight plan.

If the weather looks good and the area is secure for your aerial mission, then place the flight case in a good takeoff area, remove the camera/gimbal lock, and power up the drone, Wi-Fi extender, and RC controller. Get your shaded flight deck assembled with the iPad and controller in place while the UAV is finding satellites for GPS location and navigation. After the Phantom locks onto at least seven satellites and sets "Home Point" giving you a green light and "Ready to Fly" status on your DJI Vision app, it's time to download the background map.

Swipe your finger on the iPad within the Vision app to the mini-Ground Station display and wait for the satellite picture to load. If your iPad does not have cellular, then connect Wi-Fi to your iPhone or local Wi-Fi connection to download it. I strongly recommend against flying any new area you are not familiar with or even a large area you are local to without having a good Ground Station satellite map of where you are and where the bird is.

The map background will help you be a better navigator and make quick decisions in the air for photography shots and camera angles. The most important reason to have an accurate map background is that if you do have rapid power loss and the bird goes down, you will have a mark on the map where it went down so you can retrieve it successfully. Trust me on this one; it saved my bird one time.

9. Pack-up and -out Procedure

Hopefully you had a great flight with stable air and good light, and captured ample video and stills for your client. Now it is time to shut down the UAV and get packed to head to your home office. Before you power down, make sure all video had been stopped so the last stream will sync to the Micro SD card. Then power down the Phantom, the controller, Wi-Fi extender, and the iPad. It is easy to get into a hurry upon exiting the site and leave the controller or extender still turned on and drain the batteries.

Take the time to put the camera/gimbal lock back in place before placing the Phantom in the flight case. Then store everything else in the case in the reverse order you took it all out to make sure you are forgetting nothing.

As soon as you arrive at your home office, immediately download your day's work to your workstation, so the project does not exist solely on a single media source. It would really be a bummer if you spent two hours filming in clean air at sunset getting awesome video to let it sit on the Micro SD card for a week. Then, as media cards do sometimes, the card went bad over those few days and now you cannot retrieve the aerial video and images. Leaving it to later to download becomes troublesome when dates and times get confused, and you end up overwriting video, deleting files, or even formatting the Micro SD card during a following flight after forgetting to download the previous flight's data.

10
Software and Firmware Management

The primary component of your aerial commercial infrastructure is hardware: A flying camera. It can also be classified as a flying computer with all the intelligence built into the Phantom. Combine your smart bird with the upgradeable remote controller; the iPad for FPV (First Person View) and avionics; a dedicated video editing workstation; and your smartphone for communications and app assistance; and you will have plenty of computer software, applications, and firmware to manage and maintain consistent compatibility.

1. Workstation Operating System

The home base for your aerial operation will be your video editing workstation. Not only will you massage your high-resolution photos and HD videos on this computer, but you will also use it to update your drone, remote controller (RC), iPad, and iPhone or Android phone. Due to a larger number of available applications for video and photo editing, I highly recommend making this high-end personal computer Windows-based, rather than Apple OS or any type of Linux. You will also want to keep it protected from malware and viruses, so limit the web surfing. At the same time, do not overburden the computer with obtrusive computer security suites such as Norton or McAfee. Also, limit the number of other non-aerial applications and programs.

Do not think of using an old, junky XP PC, but do not jump too far ahead with a Windows 10 computer either. Try to use a reliable version of Windows OS, which for now is still the solid Windows 7 Professional 64-bit platform. The goal here is to have a stable and robust computing environment to edit your aerial videos and upgrade your drone entities.

Finally, make sure you are making image backups onto external hard drives to preserve your work and in case you need to recover all applications.

2. Phantom Firmware

After numerous upgrades between 2014 and 2015, the Phantom 2 Plus latest firmware level as of this writing was V3.8. With each revision of Phantom firmware, the flight stability, reliability, and increased capabilities have been delivered from the DJI website via the Phantom Assistant software. One of the first steps to configuring your new Phantom 2 Vision Plus is to download and install the Phantom Assistant on your workstation and connect the supplied USB cable between the workstation and your new Phantom 2 Vision Plus. As long as the workstation has a live connection to the Internet, the Assistant software will inventory and update all the firmware components to the Phantom as necessary.

This process may take one or more power reboots of the Phantom and repetitions of the upgrade process, depending on how old the firmware is on that particular date. Follow the instructions given by the Assistant until all firmware instances are "Up To Date." Phantom firmware upgrades add value, features, and safety most every time, but occasionally bugs are introduced and problems can arise. My recommendation is to first upgrade your reserve Phantom and perform a comprehensive suite of flight and functionality tests over a period of several days. Only after flawless sorties on the latest firmware upgrade do you configure the primary UAV with that version. Remember to also upgrade the RC code and the DJI Vision App software on your iPad so to avoid any incompatibilities between devices.

3. Remote Controller Updates

The DJI Phantom Remote Controller (RC) has fewer firmware upgrades than the Phantom itself but the updates are usually deployed in tandem. Whenever you have updated your Phantom with the Vision application, connect the RC with the same USB cable but use the DJI Remote Assistant program to check for needed firmware upgrades. Just like the Vision program, the Remote Assistant will automatically download and apply the upgrade to the RC. After the firmware download and installation is complete, you may be instructed to recalibrate the RC.

Testing is highly recommended after RC firmware upgrades as well as other upgrades. Take the drone up and perform normal aerial maneuvers, and retest the "Go-Home" feature by turning off the RC completely. What you do not want is a significant version difference between your Phantom firmware level and the RC's version. That is why it is critical to check both the Vision and Remote Assistant programs on the same date.

4. iPad and iPhone Operating System Levels

The DJI Vision app runs on the iPad, and it is important to keep the most up-to-date version of that application installed. This is also true in reverse; after you upgrade the Vision App on the iPad via the Apple iTunes Store, check for the accompanying DJI Vision and Remote Assistant firmware upgrades right away. Keep your iPad at the current operating system (iOS) level, since apps can quickly become dysfunctional after iPad operating system upgrades are performed.

The same goes for your iPhone iOS level; keep it updated so you don't run into compatibility issues with apps. One problem I ran into was an issue tethering my iPhone and iPad; once, after a late night upgrade to my iPad, I was not able to connect to the Internet from my iPhone under the personal hotspot option to download the Ground Station maps for that location. This gave me a blank background in the DJI Vision app when I swiped the screen right, and I had no satellite maps. Without the map I could not set the waypoints and fly the mapping task of my aerial project. It turned out that since I only upgraded my iPad and not my iPhone, an incompatibility arose that disabled the binding capability of the iPad to the iPhone's personal hotspot. If I had upgraded them both at the same time, I would not have had that problem in the field, which cost me money in lost work.

Basically, you may have to chase operating system upgrades and round-robin app updates to get compatibility harmony for all the computing components of your aerial support systems.

11 Rules, Licenses, Exemptions, Observers, and More

1. FAA 333 Exemption Certificate of Airworthiness (COA) Process

While you are setting up your aerial photography infrastructure and learning how to fly drones, you may have to go through a painful process to become a legal commercial pilot and obtain the Federal Aviation Administration 333 Certificate of Airworthiness (COA). The cumbersome triple-3 exemption process, which basically means you're not subject to the civil FAA rule banning commercial drone activity, will probably still be in effect through early 2016 as the first release of this book becomes available on Kindle and iBooks. My hope is by the time this manuscript is in paperback in the second half of 2016, the final FAA Unmanned aircraft systems rule set will go into effect and significantly reduce the complexity of becoming an FAA-compliant commercial pilot.

Due to the foot-dragging the federal government sometimes does, the 333 process may linger on past 2016; time will tell. Until the final implementation of the FAA UAV rule set, any drone enthusiast who wishes to profit from aerial services will still have to endure the painful

333 exemption path and either hire a licensed pilot or be one himself or herself.

After flying numerous no-charge market research flights just to see what kind of services I could produce when and if I decided to go pro, the big thing I learned was that prospective aerial photography clients needed the source to be clean. In other words, they would only purchase aerial services from a drone photography company that was fully FAA compliant. This, along with the potential of a $10,000 fine per violation (if I were caught flying commercially without the COA), inspired me to pursue the FAA 333 exemption pathway for full federal commercial UAV compliance.

2. Getting the Triple-3 Ball Rolling

Hopefully you will not have to endure the ridiculous process the FAA is making the first couple of thousand commercial drone pilots suffer through, as I did in 2015. After making numerous calls to lawyers who had prepared various FAA 333 exemption requests, the prices I was quoted ranged from $5,000 to $10,000 for the lawyers to submit the paperwork for me. Luckily, I noticed that on the FAA website the feds were posting the original 333 petitions in PDF format that could be copied and pasted into MS Word. Next, I found a company that was requesting the same operations I wanted to pursue, which were real estate and construction photography. The small-business owner who had just received his FAA 333 exemption was more than happy to give me permission to use his original request as a template to compile mine.

That next week, I submitted my customized FAA 333 exemption request myself, directly to the FAA without a lawyer, and started the countdown clock. Back in April of 2015, the 333 granted count had just passed 600 with a processing time of four months. To my surprise, I received my coveted letter of the 333 exemption granting from the FAA in fewer than three months, by the summer of 2015 (see Sample 2). Along with the exemption was the 333 waiver for aerial photography and generic data acquisition flights for less than 200 feet. (See www.faa.gov/uas/legislative_programs/section_333/how_to_file_a_petition/ for more information on what is required.)

3. Certified FAA Pilot Credentials

The blanket Certificate of Authorization (COA) also reduced the requirement of the pilot to hold just a Sport Pilot certification rather than a Private Pilot rating, which does not require as many flight hours.

Sample 2
FAA Letter

U.S. Department of Transportation
Federal Aviation Administration

800 Independence Ave., S.W.
Washington, D.C. 20591

July 23, 2015

Exemption No. 12126
Regulatory Docket No. FAA-2015-1594

Mr. John David Deans
Deans Consulting LLC
dba Central Texas Drones
6206 Ganske Road
Burton, TX 77835

Dear Mr. Deans:

This letter is to inform you that we have granted your request for exemption. It transmits our decision, explains its basis, and gives you the conditions and limitations of the exemption, including the date it ends.

By letter dated April 30, 2015, you petitioned the Federal Aviation Administration (FAA) on behalf of Deans Consulting LLC dba Central Texas Drones (hereinafter petitioner or operator) for an exemption. The petitioner requested to operate an unmanned aircraft system (UAS) to conduct aerial videography and photography.

Since I got my private pilot's license when I was 19, this set me up to be a hot commodity on the UAV market. The problem was that I had not flown a real airplane in more than 20 years!

Many of the 333 exemption holders were corporations and had to hire current FAA pilots to fly their drones to remain fully compliant. I saw a great opportunity to have both the triple-3 exemption and be my own UAV pilot. So, it was back to flight school to earn by Biannual Flight Review sign off.

It took a half day of ground school, then four long flights with an instructor to get my air legs back. Gone were the days of analog avionics; now all the dials and indicators have been integrated into an iPad-sized screen on the dash in front of the left seat of the Cessna SkyCatcher high wing. My instructor was patient but thorough, making sure that I got caught up to the new digital avionics and could actually fly the pattern, communicate with the tower, and land the plane safely.

Next I had to attain my Third Class Medical Certificate, and verify my eyeballs were working properly. Also, my physical paper license was so old that the pilot number was my social security number, so I had to request a new nine-digit plastic license which took several weeks.

The most frustrating requirement was that I had to actually get FAA identification numbers for my DJI Phantom quadcopters. That is right; just like the big Boeing 777 November numbers on the tail, I had to fill out a very specific carbon-copy form, an affidavit of ownership, and a letter specifying the layout of my corporation. It took several iterations of those documents, but I finally received the N-numbers that are tied to the serial numbers of the Phantoms. With a label maker, in the largest font possible, I placed my FAA-registered numbers on my two drones.

The samples are some of the documents I had to send into the FAA Oklahoma office (where the branch we're dealing with is) to get the aircraft registration numbers for my drones (see Samples 3, 4, and 5).

After I was officially FAA sanctioned, I had a 333 exemption number, a docket number, a new private pilot's number, and an N number for both of my Phantom 2s. Now I get to file an addendum to be authorized to fly a DJI Phantom 3, wait a few months, and then file for a third N-number for that new bird.

A couple of years from now, commercial drone pilots will be known as members of one or more of the early groups: 1) Outlaw Moneymaking Droners, 2) Triple-3 exemptions, or 3) Licensed UAV commercial pilots. Some of us will be in one or more of those original groups.

4. FAA Rules You Need to Follow after Certification

4.1 Filing FAA UAV Flight Notices to Airmen (NOTAM)

Perhaps the most time consuming reoccurring requirement being a commercial drone pilot is filing the Notice to Airmen (NOTAM) with the FAA's regional office before every flight. Only the moneymaking droners have to do this, and it is a royal pain. The FAA's reason for this is to notify other pilots of the UAV activity in specific areas at certain times. This does not reserve our airspace during those time windows, but it does give us some cover of liability or defense of us being up there if a low flying helicopter hits our drone at 175 feet in unrestricted airspace.

As stated in your COA that came with your FAA 333 exemption, you are required to phone in the flight event information, which is a subset

Sample 3
FAA Certification

GPO U.S. GPO: 2013-746-410

FORM APPROVED
OMB No. 2120-0042

UNITED STATES OF AMERICA DEPARTMENT OF TRANSPORTATION
FEDERAL AVIATION ADMINISTRATION-MIKE MONRONEY AERONAUTICAL CENTER
AIRCRAFT REGISTRATION APPLICATION

CERT. ISSUE DATE

UNITED STATES REGISTRATION NUMBER N

AIRCRAFT MANUFACTURER & MODEL: DJI Phantom 2 Vision +

AIRCRAFT SERIAL No.: PH645266488

FOR FAA USE ONLY

TYPE OF REGISTRATION (Check One box)

☐ 1. Individual ☐ 2. Partnership ☑ 3. Corporation ☐ 4. Co-Owner ☐ 5. Government ☐ 8. Non-Citizen Corporation ☐ 9. Non-Citizen Corporation Co-Owner

NAME OR APPLICANT (Person(s) shown on evidence of ownership. If individual, give last name, first name, and middle initial.)

Deans Consulting LLC

TELEPHONE NUMBER: 979 203 1534

ADDRESS (Permanent mailing address for first applicant on list) (If P.O. Box is used, physical address must also be shown.)

Number and street: 6206 Ganske Rd

Rural Route: P.O. Box:

CITY	STATE	ZIP CODE
Burton	Texas	77835

☐ CHECK HERE IF YOU ARE ONLY REPORTING A CHANGE OF ADDRESS

ATTENTION! Read the following statement before signing this application.
This portion MUST be completed.

A false or dishonest answer to any question in this application may be grounds for punishment by fine and/or imprisonment (U.S. Code, Title 18, Sec. 1001).

CERTIFICATION

I/WE CERTIFY:

(1) That the above aircraft is owned by the undersigned applicant, who is a citizen (including corporations) of the United States.
(For voting trust, give name of trustee: ____________), or:
CHECK ONE AS APPROPRIATE:
a. ☐ A resident alien, with alien registration (Form 1-151 or Form 1-551) No. ____________
b. ☐ A non-citizen corporation organized and doing business under the laws of (state) ____________ and said aircraft is based and primarily used in the United States. Records or flight hours are available for inspection at ____________

(2) That the aircraft is not registered under the laws of any foreign country; and
(3) That legal evidence of ownership is attached or has been filed with the Federal Aviation Administration.

NOTE: If executed for co-ownership all applicants must sign. Use reverse side if necessary.

TYPE OR PRINT NAME BELOW SIGNATURE

EACH PART OF THIS APPLICATION MUST BE SIGNED IN INK.

SIGNATURE	TITLE	DATE
John David Deans (signature)	Manager	8/16/2015
John David Deans		

NOTE Pending receipt of the Certificate of Aircraft Registration, the aircraft may be operated for a period not in excess of 90 days, during which time the PINK copy of this application must be carried in the aircraft.

AC Form 8050-1 (5/12) (NSN 0052-00-628-9007)

of a formal flight plan. This must be done at least 24 hours before the flight but no more than 72 hours ahead of the event. In other words, call it in one to three days before you take off.

4.2 Pre-call preparation

Before you call the FAA at 877-487-6867, you will want to have your aerial ducks in a row. First, get a good address for the flight area from your client and look it up on Google Earth. With that location on your

Sample 4
Affidavit of Ownership

Paperwork Reduction Act Statement: The information collected on this form is necessary to ensure applicant eligibility. The information is used to determine that the applicant meets the necessary qualifications as owner of an amateur built aircraft. We estimate that it will take approximately 30 minutes to complete the form. The information collection is required to obtain a benefit. The information collected becomes part of the aircraft registration system. Please note that an agency may not conduct or sponsor, and a person is not required to respond to, a collection of information unless it displays a currently valid OMB control number. **OMB 2120-0042.**

Comments covering the accuracy of this burden and suggestions for reducing the burden should be directed to the FAA at 800 Independence Avenue SW, Washington, DC 20591. ATTN: Information Collection Clearance Officer, AES-200.

AFFIDAVIT OF OWNERSHIP FOR AMATEUR-BUILT AND OTHER NON-TYPE CERTIFICATED AIRCRAFT

(does not include light-sport)

U. S. Identification: Deans Consulting LLC

Name of Amateur / Non TC'd builder: DJI Industries

Model: Phantom 2 Vision + Serial Number: PH645266488

Class (airplane, rotorcraft, glider, weight shift control, powered-parachute, etc.): Rotorcraft

Type of Engine Installed (reciprocating, turboprop, 2 or 4 cycle, electric, etc.): Electric

Manufacturer, Model and Serial Number of each Engine Installed: No Serial Numbers are available on any of the four engines Number of Engines Installed: Four

Built for Land or Sea Operation: Land Number of Seats: 0

MUST CHECK ONE

☐ More than 50% of the above-described aircraft was built from miscellaneous parts and I am the owner. (This option is for aircraft eligible for amateur-built certification.)

☐ More than 50% of the above-described aircraft was built from a kit (prefabricated parts) and I am the owner. The bill of sale from the kit manufacturer is attached. (This option is for aircraft eligible for amateur-built certification.)

☑ I certify that the above-described aircraft is a newly built non-type certificated aircraft and is not currently registered in another country. (This option is for aircraft eligible for experimental certification other than amateur-built.)

☐ I certify that the above-described aircraft is a previously built (used) non-type certificated aircraft and is not currently registered in another country. (This option is for aircraft eligible for experimental certification other than amateur-built certification.)
- ☐ Evidence of ownership from the aircraft builder through any intervening owners is attached (chain of ownership).
- ☐ Unable to obtain complete chain of ownership. Statement as to ownership history and whereabouts of aircraft is attached.

Name of Owner: Deans Consulting LLC

Signature of Owner: John David Deans Title of Signer (If Appropriate): Manager

Address: 6206 Gauske Road

City: Burton State: Texas Zip: 77835

Telephone: 9792031534

Notary Public:

State of: ________ County of: ________

Subscribed and sworn to before me this ________ day of ________, ________

My Commission Expires: ________

____

(Signature of Notary Public)

AC Form 8050-88 (09/10) Supersedes Previous Editions

Sample 5
November Registration

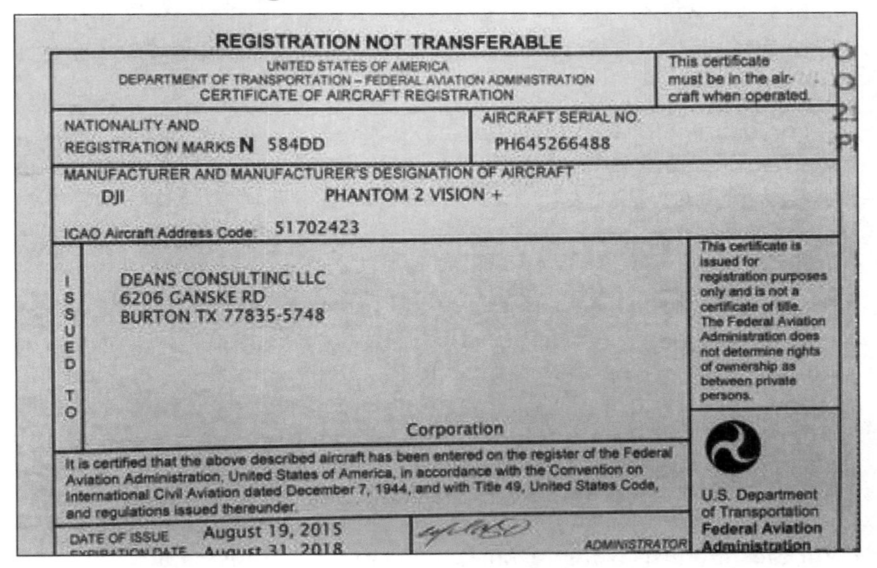

REGISTRATION NOT TRANSFERABLE

UNITED STATES OF AMERICA
DEPARTMENT OF TRANSPORTATION – FEDERAL AVIATION ADMINISTRATION
CERTIFICATE OF AIRCRAFT REGISTRATION

This certificate must be in the aircraft when operated.

NATIONALITY AND REGISTRATION MARKS N 584DD

AIRCRAFT SERIAL NO. PH645266488

MANUFACTURER AND MANUFACTURER'S DESIGNATION OF AIRCRAFT
DJI PHANTOM 2 VISION +

ICAO Aircraft Address Code: 51702423

ISSUED TO

DEANS CONSULTING LLC
6206 GANSKE RD
BURTON TX 77835-5748

Corporation

This certificate is issued for registration purposes only and is not a certificate of title. The Federal Aviation Administration does not determine rights of ownership as between private persons.

It is certified that the above described aircraft has been entered on the register of the Federal Aviation Administration, United States of America, in accordance with the Convention on International Civil Aviation dated December 7, 1944, and with Title 49, United States Code, and regulations issued thereunder.

DATE OF ISSUE August 19, 2015
EXPIRATION DATE August 31, 2018

ADMINISTRATOR

U.S. Department of Transportation
Federal Aviation Administration

screen, get out your FAA sectional map and see what airports and other airspace restrictions are in place near your prospective UAV area.

Airspace knowledge and awareness is critical when dealing with the FAA, and this is especially true when you are about to file the NOTAM.

In your COA, it specifies that you must avoid flying your UAV within five miles of an airport with a control tower, three miles of a non-tower airport that has ILS facilities, and two miles of a non-tower and non-ILS airport. If your target area does lie within those boundaries, do not even try to file for the NOTAM. You will need to file for a special COA with time-consuming multi-departmental FAA approvals all the way up to Washington, DC. It can be done, but be prepared to work that paper train for several weeks before the flight.

After you have verified you are outside the 5/3/2 airport area restrictions, then you can attain the latitude/longitude coordinates of your UAV flight area along with the date and time you plan to fly. Do not be in a hurry, and plan to spend 10 to 20 minutes on the phone with the FAA flight-planning personnel. Make sure you ask for the NOTAM number for your documentation.

Hopefully, by the time this book hits paperback, there will be an online NOTAM filing procedure or even an official FAA app for smartphones that will enable us to get automatically cleared to fly without having to talk to a human. Sometimes, that NOTAM filing takes longer than my actual UAV flight.

For more information about airspace classification go to https://www.faasafety.gov/gslac/ALC/course_content.aspx?cID=42&sID=505&preview=true.

4.3 NOTAM documentation

Keep a good spreadsheet either in MS Excel or Google Sheets online to document your UAV flight activity. For one reason, your FAA COA requires monthly reports of specific flight information. I include the NOTAM number in my UAV flight log spreadsheet just in case of an incident or a call from the FAA weeks or months later about a report received on a flight of mine.

Similar to a private pilot's log book, our UAV documentation and flight logs are just as important. Odds are it will only be after an incident like an accident or some neighbor calling you in to the authorities when you will be asked to show your drone flight log and commercial authorization documentation. It will be too late then to update it or even create it, so keep up with it on a flight-by-flight basis.

5. Observer Management

As a commercial drone pilot, the FAA mandates you use an observer during UAV flight operations. This can add costs to your deliverable, but to be compliant with the FAA 333 exemption requirements, you need to make arrangements to have a second pair of eyes on the drone in the sky. The original FAA 333 exemption document had more than 32 requirements I had to comply with and a large number of them seemed ridiculous. With the 200-foot COA overriding some of the requirements, the one that still holds strong is the use of an observer.

Even before I was interested in flying drones for money, I used my teenage daughters as observers just so I would not do something stupid with my bird up in the sky while I was trying to figure things out with the RC or the DJI Vision app running on my iPad. Having someone else next to you whose focus is on the aerial status of your UAV can be extremely helpful. This goes double for when that observer is younger and has better eyesight. Since I'm in my early 50s, trying to spot a white Phantom at 300 feet up against the background of white clouds can be quite difficult. Numerous times I would ask my 13-year-old daughter, "You got it? I can't see it anymore!" She would reply, "Yeah Dad, just to the west at ten o'clock."

5.1 The observer's tools

Have your observer install similar weather apps on his or her smartphone for monitoring dynamic weather conditions. Have your observer scan the horizon periodically to watch for developing thunderstorms or fast-moving fronts. While at it, he or she can keep an eye out for flocks of birds, aggressive hawks and eagles, and dangerous, low-flying helicopters.

The observer can also be sporting sunglasses and binoculars for long-distance position verification. This can be quite helpful on sunny days with flight vectors into the brightness or the white backdrop of clouds that match the Phantom. Another good accessory for you and the observer to have is a pair of walkie-talkies. There may be sites that require different viewpoints and lines of sight which greatly separate the pilot from the observer. Sometimes the pilot is tied to the vehicle due to power or communications requirements.

Finally, nothing beats a good old analog magnetic compass on a lanyard around your observer's neck. If the worst happens and both of you lose signal and visual sight of the UAV, the observer can say, "Last visual vector was 270 degrees at 100 feet," so that you can attempt to retrieve it.

5.2 The observer's duties

Besides observing the drone and the flight, the observer can be a big help carrying items to and from your vehicle. Time is money, so you want to get the bird airborne gathering profitable footage, wasting very little time on the ground prepping or shutting down for the day. Put him or her to work physically, checking the site to make sure no equipment or accessories get left behind.

As mentioned before, site security can become an issue in high-traffic areas and potentially hostile environments such as suburban zones. Prepare your observer to watch your back for the curiosity seeker all the way up to the dangerous street thug who may want to mug you for your $1,500 UAV. There could be law enforcement personnel that show up with simple questions about your drone or question your right to fly in the local airspace. Have your observer act as a temporary buffer from all distractions and a warning bell for any potential aerial show stoppers.

Brief your observer on the aerial issues to keep an eye out for the following:

1. Ground-based obstructions such as trees, buildings, towers, guide wires, etc.
2. Onlookers who may be in the way of takeoffs, landings, or pilot operations.
3. Ground vehicular traffic that may block LZs, pilot safety, or line of sight.
4. Large groups of small birds, or single birds of prey that may attack the UAV.
5. Fast-developing weather conditions that could produce rain, snow, hail, or high winds.
6. Deteriorating atmospheric situations that may quickly decrease visual capabilities.
7. Safety of the pilot: Keep him or her from walking into ant beds, dropoffs, and other hazards.
8. Security of the LZ: Keep a lookout for thugs and thieves and other physical threats.
9. Listen for incoming manned aircraft such as low-flying, small planes or helicopters.
10. Watch for all potentially hazardous air traffic; manned or other UAVs.
11. Aid the pilot in watching the time, spare battery consumption, and power levels.
12. Cover the pilot's communication needs while the drone is in the air; work the cell phone, etc.

5.3 Finding aerial observers

Sometimes you may have difficulty obtaining the required observer, but there are some good alternatives and quick solutions that can cover this requirement. My first source for observers are my teenage daughters, and then my buddies. The family group can be compensated with small amounts of cash and the second with beer after the bird has landed safely.

Another good alternative observer is the client who is paying you for the aerial service. I have found that a large percentage of clients want to be present during the flight just for curiosity's sake, and they wish to play the role of "director" so the correct things are filmed and the negative aspects of any area are avoided. Ask the client if he or she would be comfortable serving the role as an official UAV aerial observer. The vast majority will jump at the chance, which will give you a free and energetic observer with a vested interest in the process and outcome.

There have been a few times where my UAV attracted people who then came over while I was piloting by myself to ask questions. A couple of times this was useful since I immediately put them to use as observers. After I gave them a quick briefing on how to relay the UAV's position to me and the key points of drone observation, they became beneficial to the flight. As long as it was not a multi-battery extended session, most were quite happy to be of service and enjoyed participating in the UAV activity.

These helter-skelter methods of obtaining observers can become unreliable and inconsistent when your business activity increases. It will be at that point that hiring a young person will become necessary. Finding someone during the summer will not be difficult, but during the school year may be a challenge which may force you to up the compensation level.

However you obtain your observer, make sure he or she has sufficient eyesight. In other words, do not pick some kid with thick glasses who cannot make out an object more than 100 yards away. Also steer clear of people who don't seem to be able to focus or constantly check their cell phones to view their texts or Facebook updates. Your observer may save your bird or prevent a catastrophic aerial event if he or she can focus and notice the airborne threat before it becomes a serious problem, so be picky when choosing.

12 Learn Aerial Photography before You Start Charging Money

This chapter will cover what you need to know to get started developing your aerial photography craft. Your amazing array of hardware, software, and optics can deliver incredible images, but the pilot not only needs to fly a productive aerial mission: He or she also must have the end-product quality in mind from start to finish. In other words, the photos and videos cannot be blurry, shaky, filmed in poor light, saved in low resolution, framed improperly, or contain numerous other blunders that amateur photographers make right after they pick up their new cameras.

Being a commercial drone pilot is also about being a professional photographer with a flying camera. Do not totally rely on the advanced technology built into your Phantom. You need to fly the UAV in a manner that will enable the bird's camera to capture the environment in a pleasing style to make your clients happy. The goal here is to film, edit, and produce stunning high-resolution photos and breath-taking HD video.

1. Video- or Photo-centric Flights

Before you lift off, you need to know what and how you are going to film. Different projects require various pictorial angles and lighting along with specific UAV velocities and aerial elevations. Well before all those technicalities you have a choice of two simple options: video or photo.

Projects such as 2D and 3D mapping are, specifically, numerous still photos taken in series so they can be stitched together to form a seamless mosaic. Many event aerial sessions will include mostly video recordings to capture actions, such as races and surfing sessions. Other commercial endeavors such as real estate projects require a combination of HD video and high-resolution still photographs.

It will usually be up to you and the client to predetermine the ratio of video and photo to be captured. Once you are airborne, try to keep that percentage in mind. For example, when I am filming a 50-acre ranch for a real estate agent, I try to get a 50/50 mix of video and photos. For flight efficiency I try to capture a few photos after every leg of video flight. In other words, if I am running a left to right sweep in front of the main house, I will take three to five snapshots at a 45-degree angle to the home. Then I'll straighten out the vector, and fly the sweep with the camera in record mode pointed towards the front of the house. At the end of the run I'll stop the video recording and put back in the 45-degree angle towards the other side of the house and take a few more photos from the opposite side. This way I am constantly toggling from video to photo and back and forth during the sortie.

2. Videos in 10- to 15-second Scenes

When people are watching a video, it seems like they are not able to focus on any fixed scene for longer than 15 seconds. I first learned that when showing my initial aerial videos; when they were without significant change or activity for ten seconds, the viewers would look away and start talking to me. Only when scenes completely changed every 10 to 15 seconds did they hold their attention without looking away or talking about the video.

Noticing that low-attention span, I started keeping each scene ten seconds or less, and the whole video composition less than three minutes total. With that compilation of 18 scenes that were around ten seconds each, I could hold the viewer's attention the whole three minutes of video viewing time. I knew I filmed and edited a winner if the viewer never looked away from the screen, and said words and phrases like: "Wow!," "This is beautiful," "It's so clear," "How did you get it so smooth?," and "I love this."

3. Daylight: Morning, Noon, or Evening?

During a photography class I took years ago, the instructor told me, "It is all in the light," and "Lighting is everything."

This is true for aerial photography too. It is not enough to have 100 feet of elevation and the camera pointed down at a 45-degree angle. You have to match the feature of elevation with camera stability and proper lighting techniques.

Watch out for harsh lighting, such as morning glare where the sun is reflecting off all the wet surfaces from the dew that fell the night before. The same goes for the midday sun hitting large amounts of concrete, causing overexposure from the brightness.

Late-evening shots can be nice, as long as the shadows do not overcome your subject's composure making them too dark. Flying photography sessions are challenging due to the limited time windows, increasing shadow effect, and the hour when filming has to stop because the light is completely insufficient.

I tend to schedule my flights mid-morning or late afternoon, rather than in the extremes of sunrise, high noon, and sunset. Sunlight flicker can be especially noticeable when flying at a 90-degree angle to the morning sun, which causes a rotor visual effect to the side of the video.

4. No Drone in Your Drone Video

The last thing a client wants to see is part of the drone in the frame. In other words, make sure your drone is not in your drone video.

One of the biggest turnoffs is seeing propeller guards in the video or photography. This is like seeing the training wheels on your supposedly professional UAV. Propeller guards are for careless kids with a new $300 toy drone given to them by a dad who expects them to crash it before the first hour is done. As a commercial UAV pilot, your props should never hit anything but clean air. If you need guards, you are not ready to be a professional drone pilot. That sounds rough, but if you have the guards on now, take them off so they will not dip into the view when quickly moving forward.

The landing gear can also creep into the video from the sides when a pilot aggressively flies sideways. This can be avoided by gently pushing the right stick on the RC to the left or right. Even more caution needs to be taken in high crosswinds, since the bird will want to arc further sideways to fight the oncoming wind. You may have to travel sideways at a slower speed to prevent landing gear from showing up in your frame.

5. Smoothness Is Everything in Aerial Video

When clients are paying for aerial video, they want to see smooth transitions and not jerky turns, stops, and elevation changes. The pitch of the camera on the gimbal needs to be fixed for that scene or transitioned very slowly, to prevent any sudden camera angle changes.

The key to aerial video smoothness is pre-positioning the bird to make a planned vector recorded scene. What that means is that you want to film ten-second scenes of numerous views of the target. For example, I'll set up the UAV at 75 feet high, to the far left in front of a house for sale and have the drone's camera angle set up for it to be centered on the house, just when the center of left to right sweep vector takes it to directly in front of the house. This way there are no turns or camera pitch changes that make the video jumpy.

The next scene would be after I make a non-recorded 90-degree turn and get the side sweep vector scene set up to film the right side of the home. That way I just have to start recording and hold the right stick to the right, and the bird will fly right smoothly.

Each leg of the recorded tour can be put together in the editing session with the "fade" transition effect, so all the smooth scenes are seamlessly daisy-chained to provide a suite of ten-second video bites totaling around three minutes.

6. Stay in Motion without Repeats

Each ten-second scene should be in motion. Hardly ever should you be recording video with the UAV in a static position. If the bird is recording video, the bird should be in smooth motion either flying sideways, forward, reverse, up, or down.

In section 7, I have listed video recording maneuvers that are flown in order so none of them are skipped or repeated. (Establish your own order to ensure you never skip or repeat important maneuvers.) Of course it is easy enough to edit out flight scene repetitions but impossible to insert scenes you forgot to fly. Avoid shots with the Phantom rapidly descending since that is the least steady maneuver for the 3-axis gimbal to manage.

7. Basic Aerial Video Recording Maneuvers

Here are some rudimentary aerial video recording maneuvers, good for filming a static target such as a house or construction site (set your own starting elevations well above any obstructions). See the download kit for example videos.

1. Forward: Fly forward toward the target focal point at medium to high speed without prop visibility in the frame. Keep about 10 to 15 percent of the sky and horizon at the top of the frame with a 30-degree pitch of the camera angle.
2. Riser: Rise from near ground level to the desired elevation with no lateral movement. This gives a nice elevator effect.
3. Reverse: Fly straight back from the target focal point at medium to high speed without prop visibility in the frame. Keep about 10 to 15 percent of the sky and horizon at the top of the frame with a 30-degree pitch of the camera angle.
4. Sweep Level: Left/Right or Right/Left sideways flight at 75 feet elevation with the camera pitched with just 20 percent sky high in the picture and 90 degrees to flight path.
5. Scan Level: The UAV's position remains static at a fixed elevation while the drone starts a right turn 90 degrees, or a slow right spin from the left-most position in reference to the target. The opposite of the maneuver is starting from the right-most position and making a slow scanning turn to the left without changing the bird's horizontal plain position.

8. Intermediate Level Aerial Video Recording Maneuvers

This is a step up in aircraft control requirements. The list below requires a combination of both direction of flight and changes in elevation. See the download kit for example videos of my bird doing these maneuvers.

1. Forward Rising: Same as forward, just consistent rise in altitude. Start from about 20 feet AGL and rise up smoothly.
2. Sweep Rising: Left/Right or Right/Left sideways flight starting at 50 feet elevation then rising to 100 feet by the end of the sweep with the camera pitched with just 20 percent sky high in the picture and 90 degrees to flight path. Moderate to rapid increases in altitude are fine as long as you transition smoothly.

3. Sweep Lowering: Left/Right or Right/Left sideways flight starting at 100 feet elevation then lowering to 50 feet by the end of the sweep with the camera pitched, with just 20 percent sky high in the picture and 90 degrees to flight path. Descending too fast makes for a bumpy video so decrease the altitude gradually.
4. Rotate: Same as Scan Level but rotating a full 180, 270, or 360 degrees from the same elevation and horizontal position.

9. Advanced Aerial Video Recording Maneuvers

These flight techniques below may take some practice but the video produced can be spectacular. Transitional smoothness is the key, with no jerking motions. See the download kit for example videos of my drone doing these maneuvers.

1. Back up and out: A great finale is to start low and close then give pressure to the up and back controls to bring the bird back and up at a 45-degree ascent from as low as possible to the current limit of 200 feet under FAA 333 Exemption COA.
2. 90-Degree Overhead: Point the camera straight down from high above at 100 to 200 feet and take a moving Google Earth video.
3. 90-Degree Roll In: Utilizing the camera pitch while recording gives an interesting effect and can add a dynamic feel to your video rolling from looking forward to straight down.
4. 90-Degree Roll Out: This is the reverse of 90-Degree Roll In; by bringing the camera pitch from straight down back up to just a 20 to 30 degree pitch with just 10 to 20 percent sky and horizon showing.
5. 90-Degree Rotate: A build on the 90-Degree Overhead by smoothly rotating the bird maintaining the same horizontal and vertical position.
6. Manual Circle: This is the most difficult maneuver with a Phantom 2 Vision Plus since it relies only on smooth piloting skills. The UAV stays level anywhere from 50 to 150 feet and makes a circle around the target with the camera pointed at the center during the entire maneuver. Consistent and steady pressure on side flight with opposing turn is needed to make a smooth

360-degree flight path while keeping the bird pointed at the center of the flight's circle.

7. Road Race Mode: Picture your UAV flying just above a road racer turning and weaving on the serpentine road course. The goal here is to turn the drone while moving forward just as smoothly as a road racer would. The camera should be doing smooth pans as you bank the bird left and right.

10. Intelligent Navigation Mode Enabled Maneuvers

Here are a couple of professional level maneuvers made possible by the new Intelligent Navigation Mode (INM) within the DJI GO app coupled with the Phantom 3 or Inspire 1. See the download kit for example videos.

1. Auto-circle or Point of Interest: The INM offers the Point of Interest mode which enables you to pick a target and have the UAV in autopilot circling the "Point of Interest" in a specified radius, elevation, and velocity. This makes for a perfect arc nearly impossible to perform consistently by hand.

2. Course Lock: Another INM mode enables the pilot to establish the forward vector and while moving forward, he or she can move the right and left turn stick panning the scene without changing the drone's course vector.

Many more of these types of maneuvers can be documented and planned by mixing elevation changes and slight camera pitch changes. It all depends on your piloting and photography experience. Mix and match these aerial filming maneuvers and develop your own flight-scene list.

The most important point is to maintain the smoothness of the shot. Do not get too caught up in fancy flying if it makes the video too jumpy or shaky. Remember, the goal is to blow the client away with aerial imaging wonders and not make the person airsick.

11. Use Reverse Play on Static Scene for Precision Fly-in Effect

One trick I use for an awesome effect that makes you appear like the best UAV pilot ever can be done in the editing room. Get the bird set up at around 10 feet high directly in front of the target, like a front door to a home, with the camera pointed directly at it. Next, fly back and up at a 30-degree angle, smoothly out to 200 feet up and several hundred feet back as a fade-away video scene. This only works in a static environment without any moving vehicles or living things in the recording.

The reason for this is on the editing station you will use the "Super Tool" in PowerDirector and play the video in reverse. This gives the effect of a perfectly flown, smooth downward vector from high and out to the perfect location in front of that target door, with no turns or sudden elevation changes. You can even speed it up a little for a dramatic effect as long as trees blowing in the wind or other moving objects do not show the velocity enhancement.

12. The Invisible UAV Pilot

During the filming of your aerial video and picture taking, never should you, the pilot, be in the frame. The goal is for you to be the invisible UAV pilot to give the illusion of the flying camera being truly unmanned and providing an independent eye in the sky.

This is done by knowing the angle of the camera and avoiding its view of coverage. Basically, you and your observer will be hiding from the UAV's recording frame by positioning yourselves under trees, behind bushes, or just around the corner of buildings and vehicles. Another way is to stay behind, to the side, or just under the bird.

It is OK if you are in the video during the very start or end of the maneuver; as long as you remember to edit, you can remove your aerial crew by cropping the photo or chopping off the start or end of the video. What you really don't want is a great shot of the target scene with you and your UAV flight deck staring up at the camera. Keep yourself invisible so the client can enjoy the product without the distraction of the person who made it.

13. Avoid Drone Shadow in Your Videos

You will want moving and still aerial images without obvious indicators that all of it was filmed with a quadcopter. It can be painfully evident when the shadow of the UAV is visible in the picture or video. Be wary when you are flying close to the ground or near a building with the sun behind the bird. The wrong angle can cast an annoying shadow in the video which becomes extremely distracting to the viewer.

Be mindful of the sun's angle to the camera's view. In time and with experience you will learn under what conditions this happens so you can avoid it from the start. Another way is to keep a sharp eye on the live FPV screen during low-altitude shots and while flying close to buildings and other vertical surfaces.

You want your UAV to record in stealth mode to deliver a subject-centric photography experience.

Pilot in Shot

Drone Shadow

13 Learn 2D Mapping

The primary use for two-dimensional (2D) mapping is to have an up-to-date aerial image of a large area of land or property development. We can create a single large photograph by composing a mosaic of aerial images through a process called "stitching." The 2D mapping is especially needed in the construction market since the vast majority of Google Earth views will show an empty field where there has actually been extensive dirt work and framing of foundations over the previous few months.

Google Earth is one of the most useful applications on the Internet today. Numerous times I have put it to work finding directions, viewing potential vacation spots, and taking virtual aerial tours of locations all over the world. Though the resolution has increased over the years, there are some locations that are not so clear. The other downside of Google Earth is that less active areas are not quite up to date. Some aerial images of rural properties can be a year old, or more.

Bring on UAVs to solve both of those problems. Construction clients will pay dearly for periodic Google Earth-like, 2D high-resolution images from a 90-degree overhead angle in a single large JPG file. With a high-quality camera integrated into the Phantom on a rotating gimbal, this is possible with the right technique and software.

1. Attain Property Line Info and Last Google Earth View

First you need to know your target mapping area and its boundaries. I use our county's tax appraisal website and obtain an image showing the property line in bright green so I can plan my flights. Try to use boundary reference points such as existing roads, tree lines, and waterways like creeks and bayous to help you in the field. Set the waypoints under the Ground Station utility in the DJI Vision app on your iPad.

If the target property to be mapped is smaller than ten acres, it should be able to be flown and photographed in a single-battery flight as long as the total distance flown is less than two miles. Any larger area will take multiple sorties and will require you to lay out and plan multiple sections of flight grids and how they will overlap sufficiently.

2. Mapping Flight Grid

Let's start with a single flight grid mapping a ten-acre property with existing static features. This means the image on your Ground Station background matches the current view from the air at the time of your flight. After you have obtained the perimeter from either the client or your tax appraisal website, set waypoints in a grid configuration at an elevation of 200 feet, with each stitch leg of flight 200 feet apart.

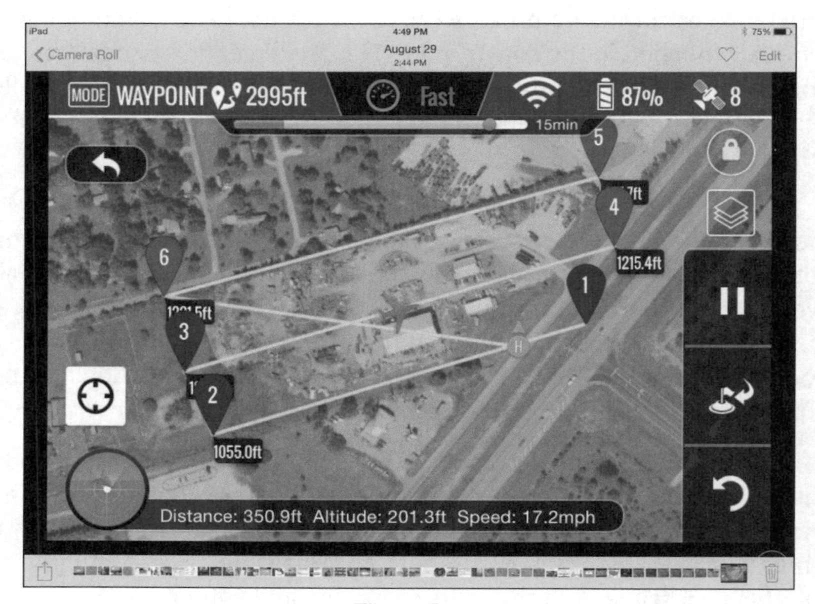

Flight Grid

This is called the 200 x 200 mapping grid which covers most areas smaller than 100 acres quite well. Larger 2D mapping projects will require a higher elevation, but since our FAA COA limits us to only 200 feet, we will utilize the 200 x 200 "rule of thumb" for aerial mapping. When future COAs allow us higher altitudes, we can bump up to the 400 x 300 grid which is 400 feet AGL with 300 feet of leg (flight segment) separation to help us cover more ground. Keep in mind, the higher we fly the lower the resolution will be after the final image is produced.

3. Static Environment Is Critical

The secret to successful aerial mapping is having a static environment to photograph and a stable platform to take those photos. We use software to gather these images, match them based on the overlays, and create one large contiguous picture that can be several hundred megapixels in size.

For the images to be taken properly so the stitching process will work, we need 20 to 40 percent overlap, and for nothing to change on the ground while we are filming. This means no vehicle or large animal movement during the aerial image capturing process. We also have to have minimal to no wind affecting our flight. I only perform 2D mapping flights with less than five mph winds and insist on no activity on the ground. The last thing you want is to finish the flight, download the captured images at home, and find out not enough images will stitch due to a truck moving throughout. Just as bad can be wind pushing your UAV into a crabbed angle of flight (where you're fighting wind), skewing your camera view, which can also have a negative effect on the stitching software's ability to match up the images.

Simply put, if you are planning a 2D mapping flight, make sure you do it on a dead-wind day with no activity planned for that area. I have had to shoot some mapping projects on a Sunday when business was closed to have a static environment, and was praying for no wind during that same time frame. Sometimes the low-wind opportunity and the lack of activity for 2D mapping is difficult to schedule, but that is why we can charge more for that service.

4. Set Camera Mapping Mode

Once your grid is laid out, hit the DONE button and verify all your waypoints are at a consistent elevation of 200 feet (plus or minus 5 feet) and your speed is set to FAST. Then toggle back to the main camera screen to make some changes there before your flight. First set the camera pitch to straight down at a 90-degree angle to the flight. Then, go into the options and set the camera snapshot to timed, with three seconds between shots.

Exit out of the camera mode and swipe back to the Ground Station and hit GO to begin mapping the flight. Just before it arrives at the first waypoint, which is the start of the grid, remember to go back to the camera screen and start the photo capture by hitting the camera button. You should start seeing the shutter take pictures every three seconds, with the screen going dark every few seconds.

Watch the UAV physically, monitor the battery level, and view the progress of the flight to the waypoints on your iPad all at the same time. When the drone arrives at the final waypoint and turns to come home, stop the picture capture to make it easier to process later. Just before the bird arrives home, take a screen capture of the grid-flight pattern on the Ground Station display for reference to create the next grid flight, if you have to perform additional sorties to capture large areas. If an additional grid is needed, create it based on a 20 percent overlap with the first grid previously flown. Swap in a fresh battery, and restart the process.

5. 2D Map Stitching

Upon returning to your home office, download the raw-mapping images to your workstation and Defish them if necessary, to remove the wide-angle fisheye effect with the DeFishr program we discussed earlier. Then, download and install the free Microsoft program Image Composition Editor (ICE) to combine the numerous straight-down photos your UAV captured.

Hopefully you captured the images at the straight up and down camera pitch, with no wind affecting the flight path, with no activity on the ground, with the proper overlap, at the right UAV velocity, and at a consistent altitude. If everything went just as you planned it and your Defishing process is complete, you should be able to drag and drop the images into ICE and watch it stitch together a single, large, flawless image that puts Google Earth to shame.

6. KLONK Software to Measure 2D Entities

Once you have a high-resolution 2D mapped image, it may be time to analyze it by taking measurements of certain features. An excellent, affordable application to perform this duty is from KLONK and is called the Image Measurement tool (www.ImageMeasurement.com). The professional version of this Windows-based program costs $35 a month and can be a very valuable tool for you and your client if recurring 2D mapping flights are in order.

You can import your high-resolution mapped image into the Image Measurement application. After setting the calibration using an object of a known size, you can then take measurements of length, circumference, radius, and area of things on the ground. This can be quite helpful to clients who have to quantify amounts on the ground or even take inventory on their property.

2D Mapping Image Measurement

The 2D mapping with UAVs can be difficult due to UAV configuration and flight condition restrictions along with the advanced software applications necessary. However, once we get the processes down, it can be a great asset and a profitable skill set to offer.

14 Your Deliverables: Develop Your Package of HD Video and Photos

The deliverable to your clients will be a package of aerial video and photos from the flight. This bundle will represent your company, your skill level, and your value, so deliver a quality product to every customer every time. Standardizing your deliverable will be important so you can be consistent, and efficient in assembling it after you return from the flight site.

My scenic package for real estate clients consists of a single HD video in .mp4 format running three to four minutes long with 15 to 20 ten-second scenes. Each scene transitions smoothly with a "blur" effect going directly into slight movement of the bird. Hardly ever are the UAV and camera stationary with a static scene with nothing going on within the frame. Your movie cannot be boring even if what you are shooting is.

1. Aerial Video Storytelling

Just after the deal is made for you to shoot a location, look it up on Google Earth to begin visualizing how you are going to tell a story about what this place is like with the video you are going to make. Look at the entries and exits to and from the road. See what a first-time visitor would experience and highlight those views from the air.

Note both the positives and the negatives of the site. When shooting a real estate project, plan to highlight entities such as ponds and lakes, ridges and hills, and all types of wildlife both domestic and free-roaming. Shy away from large power line towers, petroleum pipelines, burn piles, trash heaps, and discarded, large items such as junk cars and dilapidated barns. The realtor has to disclose the existence of those eyesores, but you do not need to showcase them.

The larger and more infrastructure-rich the site is, the more you need to compile a structured flow to your video. Have an opening scene entering the property from the road at a higher altitude, going forward but lowering as you proceed through the gate. Work your way around the residence, showing the property boundaries by flying sideways just outside the external fence line with the camera pointed inside the target area.

Show those positive aspects in order of importance, saving the main residence for the final scenes. Each scene should use a different maneuver such as a sweep, then scan, then rise, then rotate, then forward, etc. Try to get some sort of a 45-degree view from east, west, north, and south. Then mix in some overheads with the camera pitch at a 90-degree, straight down angle.

Finish up with the main house or the primary object as the parting shot in a reverse back away and up scene from 20 feet, rising up and fading out at 200 feet elevation, with 1,000 feet of distance away from your starting position.

2. Blend Music and Video

Your music can be an important component of your video. Know the client, so you do not have a hard rock song playing in the background for a long-time country music fan's aerial of his family's ranch. One option is to ask the customer what flavor of background melody he or she would prefer. Another option is to purchase some moderate, mellow songs from Pond5.com that feel laid back and relaxing.

Whatever song you choose, make sure it is long enough to span the complete video. What you do not want is to have to repeat the same tune or have two different melodies. I try to only purchase songs that are four to six minutes long, since I can always end them by fading out the soundtrack at the end of the video.

3. Video Titles and End Credits

Use the "Insert Text" function within the PowerDirector application to add a title to your video with at least the year, if not the full date and time. Something like "The Smith Estate — Fall 2015" will work. Make sure you fade the title in and out smoothly with no sudden appearance and removal.

Avoid too many text inserts in the middle of your video, since that can be distracting with too many things to read. That said, some clients may want pop-up captions explaining features of the property such as acreage, building specifications, and other aspects of what is being displayed.

As the last scene fades away, insert your credits showing your FAA 333 Exemption number or your UAV license along with your website and phone number. This way, you get free advertising to the viewers who just saw the aerial video and are blown away.

4. Still Aerial Photographs

During the flight, you should have taken high-resolution shots between the video scenes. If necessary, Defish those large images and place them in a folder called "High Resolution Aerial Photos." My goal is to have at least 20 of these separately taken images from around each property.

Alternatively, use the free VLC player program and take dozens of snapshots while playing the video you created. Look under the Tools/Preferences/Video/Video Snapshots and create a folder called "Photos from Video." Then play the complete aerial video while taking snapshots by clicking the small camera icon on the lower left every five seconds or so. Though these images are lower resolution than the photos shot directly from the Phantom, they are still clear enough for screensavers, social media, and digital picture frames. You will be surprised how many photos you may create while watching the aerial video, although your rate of photo creation will usually depend on the richness and action in the scene. Watch out for photos taken during the blurry transition of the scenes since that will look odd.

5. The Deliverables

Make sure you have already purchased a bundle of low-cost 4GB USB drives so you can place your deliverables on them for clients. My client USBs contain the production HD aerial mp4 video and two photo folders of "High Resolution Photos" and "Photos from Video." I then place the USB drive into an envelope with the hardcopy invoice so I can get paid as soon as they see the video and are smiling from ear to ear.

Another method to distribute your aerial deliverables is to upload them to Dropbox.com and email the client a link. This is handy if the client is in a big rush or meeting in person is logistically difficult. I prefer a face-to-face meeting to show the final product.

I also host the aerial videos on YouTube.com and Vimeo.com for two reasons: First, I can email video links to clients so they can in turn distribute their newly acquired video to their friends and family, which is great advertising for me. Second, I use those same videos in my marketing efforts unless the client explicitly requests it stay off the Internet. In those cases I explain that Google Earth already has images of their land on the web, so why not at least make it look good? They usually agree to let me share it after I explain it like that.

6. The Aerial Presentation

One of the reasons I strongly recommend presenting the final aerial product to your client face to face is that it must work right the first time. If you hand a client a USB drive and say, "Have at it," something may go wrong with the playing of the video. The client could plug that drive into an old Windows XP computer that runs like junk due to malware and bad graphics.

I do not want my high-quality aerial video to freeze, jump, and look bad due to a client's outdated hardware and polluted software. As a 35-year IT veteran, I see computer problems all the time, so I want to make sure the client does not think my video is bad when it is really his or her deficient computer.

This can be a touchy situation, so what I do is take either my laptop or, sometimes, a full 40-inch flat screen LED TV with me for the presentation. With my own hardware, I know that the first time the client sees my aerial work it will be flawless, visually and audibly. The first impression is everything, and I want no technical difficulty to ruin that for my client.

Only after he or she has seen how awesome the aerial video and photos are, will I plug the USB drive into his or her computer and re-play it to see how it works. I have had to put in some extra time on my own dime working on clients' PCs, Macs, and Smart TVs to get the video to play properly. Sometimes I'll have clients only play it on the new 60-inch Samsung LED TV hanging on the wall rather than a PC, after I explain to them that the larger the screen the better the aerial looks.

7. Your Deliverable: Your Responsibility

Once you land the drone, the job does not end. As a commercial drone pilot selling photos and video, you are also an editor, composer, and a product presenter. Put in the effort to package and consistently deliver a high-quality product of production level HD videos and a stunning array of aerial photos to your client. Also be ready to deliver them in a first impression mode, so the client is blown away by your unique talent and ability.

8. Large Print, High-resolution Aerial Photos

HD videos and high-resolution photos in digital format are the two primary deliverables you will be offering as a commercial drone pilot. A third option for generating income is to provide large format high-resolution aerial photographs printed on 36 x 24-inch or larger photo paper from a local print shop, Walmart, or even your own plotter.

Framed Aerial Photo

Having the ability to print a huge color aerial photo and then mount and frame it can be both profitable and a way to distribute more advertising of your service and product. Handing your client a USB drive with 100 photos is great and he or she will like it, but also delivering a framed three-by-two-foot aerial rendition of his or her ranch to hang on the wall at home that night will really knock his or her socks off.

8.1 Picking the right image for a suggestive sell

You could choose the best high-resolution photo of the aerial project with the clearest picture, optimum lighting, and just the right angle. From there you will be taking a risk of the client not liking your rogue choice, but it is worth the chance. Alternatively, you can have it printed on a large format photo printer then have it mounted and framed very inexpensively. Then you can simply surprise him or her with the finished product and request a decent price for it.

I choose the best image of the aerial project and put it on a USB drive. I then take it to Walmart with a photography department that has a large format printer and put it into the self-serve photo machine. A few hours later, I grab the $25 36 x 24-inch hardcopy photograph rolled up in a bag, then take it to my office to mount it on a same-sized self-adhesive foam board. After that sets for a day, I place the mounted photo in a cheap, flat, black plastic frame I also purchased from Walmart for $20.

The key technique here is to mount the photo onto the self-adhesive foam board rather than press it against a cardboard backing or utilize spray adhesive. The reason for this is to prevent an air bubbling problem weeks or even months down the line. Dry mounting would be preferred, but it jacks up the cost a bit. I suggest removing any clear plastic or cheap glass in the frame and leave the matte photo exposed to avoid glare.

My total investment when I do this is less than $50, and it takes maybe half an hour of effort. I then offer the ready-to-go photo for at least $100. At least 80 percent of the time clients will purchase the additional framed hardcopy photograph, and all is good.

For a step up, you can take it to your local Hobby Lobby or equivalent store, and have it dry mount the photo; there, you can pick out a nicer frame for the client. Just make sure you increase the price to cover your costs.

15 Step-by-Step Aerial Photography Project Process

The other chapters in this book offer a pretty good idea on how to prepare and conduct an aerial photography project. In this section, let's put to use all the previous tidbits of advice, recommendations, and techniques and walk through all the steps of a sample real estate aerial project. Use this chapter as a quick list of steps that lays out the process from marketing your service to collecting your fee.

1. Step One: Marketing and Selling

Find a local real estate broker or agent who specializes in rural properties, primarily in your area. This can be done by browsing the realty pages in the newspaper or surfing the MLS websites noting who is selling ranches, farms, and acreages.

After you pick the realtor's office with the most listings and the higher-quality properties, go to its website and identify the managing agent or broker. Now here is the hard part: You will most likely have to cold call the person. Prepare some bullet points, but not a written speech. Just make it quick and get to the point by requesting a short meeting with him or her so you can stop by and show your aerial video examples, and lay out how you are a legal FAA-endorsed UAV/drone pilot who is providing low-cost aerial showcasing for real estate marketing.

When you get the meeting, be on time and bring either a nice laptop or even a portable 32-inch flat-screen TV with an extension cord so you can put it anywhere in the office. Show a fast-moving aerial comprised of your best scenes from as many sites as possible, appearing in a random mode. This way, you can start it quickly while you are talking and trying to close the deal.

Ask the realtor to give you a test property to perform and try to set a date for filming while you are still in the office. When you get the gig, have the agent call the property owner and get permission for you to fly over the property. Explain what the deliverables are and agree on a price. Mine is around $300 for an estate that is less than 100 acres and within an hour's drive (see Chapter 19 on pricing for more information). Obtain the address and owner's name, shake hands, and inform your new client he or she will have it in a few days depending on the weather.

I operate without formal contracts but you should do what works best for you.

2. Step Two: Flight Planning

Now that you have a paid gig on your to-do list, time to perform some flight planning. First, go on Google Earth and search for the property to find the location of the nearest airport. Look for tower-controlled airports and verify the site is greater than five nautical miles. Airports without a tower but with Instrument Landing Systems (ILS) need a three-mile radius of clearance, and small airfields with neither tower nor ILS only need a two-mile no-drone zone.

After your airspace has been confirmed clear of airports, look for other obstructions such as power lines, cell phone towers, lakes, or wooded areas. Next, use the address on the county's tax appraisal website, get the property line on a map, and print it to take with you for the flight.

Research the property on Google Earth and pick out interesting and positive aspects of the property to highlight and showcase. Note any negative parts of the land to avoid filming, by either skipping those parts or adjusting your camera's pitch or flight vector. Determine if the site requires multiple sorties or if it can be flown in a single-battery flight.

Make logistical arrangements with your spotter and determine meet-up points or a pickup location so you can brief him or her before site arrival. Worst case, talk to the client or property owner to see if he or she could serve as your observer.

Check the weather and find a good day with light winds and little or no rain forecasted. Remember to file the NOTAM with the FAA at least one day before the flight, but no more than three days early. Charge all your batteries the night before then pack all necessary items into the flight case and placing it by your front door. As you are packing your flight case, make sure you do not bump a switch and turn it on, thereby draining the batteries overnight.

3. Step Three: Fly the Site

The day of the flight, keep an eye on the sky and your weather apps to monitor the wind and changes of rain. Much better to delay the shoot rather than have to do it twice due to arrival at a property with 20 mph winds blowing your bird around the site. Your optimal wind is none, but go ahead and fly as long as the wind is lower than the teens. If there is a moderate chance of rain, watch the radar to monitor thunderstorms and fronts in relation to your location of flight.

When you arrive, either drive or walk around the site gathering additional property features, looking for unseen obstructions not viewable from Google Earth, and talk to the landowner. Have your observer on duty and start your preflight procedure by setting up your shaded flight deck, downloading the satellite background for the integrated Ground Station screen, and performing a compass calibration operation.

Once your Phantom has obtained the necessary number of satellites and you see the Home Point Set then Ready to Fly, it is time to take off from your flight case. Perform the sweeps along the sides of the property while recording, and pause in the corners for high-resolution photos. Also, while in the corners of the property, slowly and smoothly turn the UAV while recording video to capture a static position scanning from left to right then right to left. Remember to toggle back to taking high-resolution photos between each video segment.

Do a couple rise-up maneuvers, at least one circle fly-around focused on the house, a few reverse and forward runs highlighting noted features of the property, and then film the exit scene with a 30-degree reverse rising up and back shot for the finale.

All the while filming and flying, you and your observer need to keep an eye and ear out for airborne threats and ground-based distractions. Situational awareness is everything while the bird is in the air.

When you drop below 50 percent main UAV battery life, it's time to wrap up that sortie and bring the drone back to you. Have it in your hand or back on the flight case before it hits 40 percent. Do not risk your bird or vulnerable entities on the ground flying it at less than 40 percent due to the risk of rapid power loss issue. If this is a multiple sortie project, swap the battery and continue flying. It's much better to fly and film too much than not enough, so be generous with your air time.

After you are sure all the property's positive features have been fully videoed and photographed, it will be time to power all devices down, pack up your flight case, and head home to process the media. Put your observer to work ensuring all equipment is retrieved.

4. Step Four: Editing Process

As soon as you arrive at your office, start the download procedure with the USB cable connecting your Phantom to the editing workstation. With all the video mp4s and the photo JPGs downloaded into the RAW directory of the client's folder, proceed with the Defishing process if you are flying a Phantom 2 Vision Plus to remove the fisheye effect. During the Defishing process, you can separate the JPGs from the mp4s by placing them in different folders.

Next, view each video mp4 and name them based on the scene. I give them names like North-Sweep-L-R.mp4 and Final-A.mp4. You will probably throw out several scenes that were aborted, too shaky, or were accidentally recorded during transitional moves. Place any non-needed mp4s in a folder called Unused rather than deleting them just in case you need to go back to one.

This may take quite a while if you have dozens of scenes to view and rename. Taking the time to do this, however, makes the editing phase much easier, since you will be able to place the videos in the proper order to help you tell the story of the aerial adventure.

Another step I perform is to capture screenshot images of the highlighted property line from the county's tax appraisal website to use in the production video. I capture the property layout from three to five elevations to give the viewer an idea where the estate is located in the county, what other land features surrounding it look like, and to enable the person to see the perimeter of the land from directly above.

After you have finished the viewing and renaming process, start up the PowerDirector 13 video editing application and start importing all the newly named scenes from the UAV flight and the few property line images from the tax appraisal website. Start out by daisy-chaining the overhead property line shots then add the drone scenes as they seem appropriate.

Use the editing features of PowerDirector 13 to trim out the boring, static, and transitional starts and ends of each scene. Try to have each scene run 10 to 15 seconds after trimming is done. This way you will have 15 to 20 scenes comprising the aerial video delivering a non-boring and continuously moving film.

After all the scenes are in place, use the transitional tool to add the "blur" or "fade" effects in between each scene so the transition from one to the next is clean, smooth, and appealing to the viewer. Then add your text at the front introducing the property and your credits (free advertising) at the end of the video.

Now that you have the full content in place, you can overlay the music track with a single song you purchased from Pond5.com or similar copyright-free musical website. Do *not* use copyrighted music from a CD you bought or an album you downloaded. We want to stay legal on both the drone usage and the music we utilize, so don't be a pirate.

View the completed video a couple of times to verify there are no awkward transitions, abrupt maneuvers, or boring parts before you cut the final mp4. After it all looks good, use the Produce tab and select the "MPEG-4 1920x1080/30p" to create the projection video. This format has worked well for me on PCs, Macs, iPads, and Smart TVs. It also delivers sharp HD format with moderate disk space requirements. Most of my three- to four-minute movies are 400MB to 500MB in size.

Play the video on Microsoft Media Player for first-level inspection and quality assurance, checking for smoothness, audio, and content. Then play it again on the VLC Player and take 50 to 100 snapshots, creating that many pictures to store in the folder named "Photos from Video."

For clients who want online access to the aerial, upload the video to YouTube.com and remember to write a neat description of the movie as well as hit the Publish button, so it will be public. You may also want to upload it to Dropbox.com so the client can download it remotely if distance is an issue.

Finally, copy the mp4 production video and the two photo folders of "High Resolution Photos" and "Photos from Video" onto the 4GB USB thumb drive that will be delivered to the customer. Just to make sure the copy was successful and the USB drive is good, take it to another computer and try to view the video and photos.

5. Step Five: Presentation to the Client

As I said before, try your best to meet with the client and present the aerial video in person, preferably on your large screen laptop or portable Smart TV. This way, the client sees it big and bright for the first time without any technical problems that distract from your great work. Plus, you will get the satisfaction of seeing your client's eyes light up and hearing him or her say, "Wow! This is awesome. I love it!"

Only after clients have seen the video and pictures on your hardware do you try to view it on their computers. Help them, if possible, to copy to their computer first, since it will play better from their hard drive compared to from the USB drive. You may have to adjust the volume, disable the mute, and configure the Windows Media Player application for the first time. I have even had to install VLC Player for clients just to get my aerial video to play on a computer.

Another good reason to personally present the final aerial product is that you can get paid immediately, compared to waiting days if not weeks for a check in the mail. Also, you can give the client some business cards while he or she is pumped up, so they will likely be distributed to friends.

16
Aerial Photography Techniques for Specific Markets

1. Marketing Specific UAV Aerial Photography Techniques

Now that you have the generic aerial process down, there are some particular methods you need to follow in certain markets. In other words, how you approach and run an aerial project at a 200-acre ranch for a real estate shoot differs from what you do at a single-acre construction site project. Each aerial market has its particulars, and you need to be aware of them so you can quickly adapt and be prepared.

2. Real Estate Aerial Techniques

Aerial projects with large land-only ranches in the hundreds of acres can be a challenge due to sheer size and accessibility issues. Google Earth planning is critical so you can break up the areas of interest into sectional flights. With these large areas to portray, you will not only be telling a story, but will be taking the viewer on a structured aerial tour. You may even need four-wheel drive all-terrain vehicles or have to hike some areas due to hostile topography. Plan it out thoroughly, bring everything you could possibly need, break it up into manageable pieces, and fly it well the first time.

The 100-acre ranch will be a more rounded project since you will have both the home and barns to showcase along with the positive features of the surrounding land. Do a half-and-half video between the land and infrastructure to give viewers a good feel of the home they are thinking of purchasing. Again, break the project into sections and move around the property during your various sorties to capture different aspects such as fields, ridges, ponds, creeks, fences, and tree lines.

Smaller, rural properties less than 20 acres can usually be done in a couple of sorties and a single battery change. Many times I can stand in one place under a tree and shoot the whole project by peeking around the trunk from under the tree's canopy. The smaller the rural property, the more chance of armed country people within shotgun range, so be extra careful during bird hunting season.

Occasionally you will get an aerial gig for a single-acre suburban property. Shoot these during the week, and not in the summer. That way, very few people will be around. Since suburban homes are very close together, you will most likely be flying over other neighbors' property lines to get the angle of view to the contracting residence. This can cause too many interactions that can become tense. Pick a time when very few people will be home or outside and get it done quickly.

No matter the size of real estate environment you are flying, give the property owner time to clear the clutter. This can be as simple as moving a trailer or a few cars. Other times it can take weeks if the person has to clear burn piles, eliminate huisache brush, or fix miles of wood fence that have fallen down over the years. I had to reshoot a range once because the client wanted to shred his 40 acres but forgot to tell me before the flight. The point is to let a site owner prepare the property so he or she is proud to show it.

3. Construction Site Aerial Techniques

One construction-site-specific issue is timing. Timing to film a concrete pour while it is happening or at least filming during construction activity is usually requested. What you do not want to film is a vacant construction site with no workers present since that gives the impression nothing is getting done. Avoid flying directly over workers since UAVs should never be above living things. Even though they will have hard hats on, try to frame people through your flying camera at a 45-degree angle from 75 feet above.

Hopefully you will have negotiated a multi-visit aerial job so you can film each stage of construction. This will require you to keep up good communications with the site supervisor so you can schedule your flights to cover the initial dirt work, which could last months on some large projects. Next, you could film the foundation preparation phase with the workers laying the rebar and foundation forms. Then, fly around on the very day while they are pouring the concrete for the foundation. Schedule a sortie during the wood framing of the building or the raising of the steel skeleton structures. Also, get an aerial just as they are putting on the roof, installing windows, and getting ready for the building to be "dried in."

Once the outside structure is complete and the rest of the work is on the inside and not visible by your airborne UAV, wait for the opportunity for the final filming of the finished project. Then, as a final edited version, you can take sections of each phase of construction and compile an aerial timeline showcasing the project from start to finish in a single video from above.

4. Law Enforcement (LE) Aerial Techniques

Law enforcement (LE) aerial projects will be few and far between, but if you develop a good relationship with a local police department or, better yet, the county Sheriff's Office, interesting aerial opportunities may come up quickly. The key with LE is to be ready at all times to respond to an officer's aerial needs. Projects could be due to a missing person in a large rural area, a drug field operation, or a dangerous SWAT mission.

The key component here is readiness. That includes your bird, the batteries, and your personal gear. I recommend keeping a set of good hiking boots in your vehicle because odds are the LZ will be on a clearing in the woods rather than a nice, open parking lot.

Owning and using a police scanner may also help in your marketing efforts. You may hear a call about a fast moving grass fire or a SWAT operation which could quickly develop into an opportunity for your UAV to help LE clients. In our small town, I have the cell numbers to the fire chief, sheriff, and chief of police so I can call to offer my airborne eye in the sky just in case they forget about me. Compared to the cost of a manned helicopter, your drone service is a bargain if you can make it on time.

The primary problem here is keeping a couple batteries charged up at all times, so you can be ready to fly in less than 15 minutes. The only real way to do that is to rotate them out and drain them with dummy flights at least once a week. For it to be worth the effort, I would need at least one flight per month.

5. Commercial Dealerships

Though a booming market right now, once a car dealership has a drone video, they probably will not be a repeat customer. If you are one of the first legit 333 droners to make the sales call on a dealership, then you will have a good chance on getting the flight. It is still worth the effort to set a meeting with the dealership manager to show your aerial portfolio. If the dealer already had an aerial performed in the past, it may not be as smooth and professional-looking as yours could be.

While you are watching TV and you see a car, boat, or RV dealership commercial and you see the fisheye effect with the horizon bent, then you have a good prospect. Cold-call and sell the management on how you can reshoot the dealership with no fisheye effect, and better angles, smoother video, and higher resolution.

Use the same flight maneuvers to highlight the vehicles on the lot and the standard large American flag waving. I also fly a close side-shot of the dealership sign, along with the site, under some good light close to sunset.

Since they will only use a 5-second video clip for their 30-second commercial, give them numerous 10-second fade away, scanning, sweeping, and rising scenes. That way they can trim them out just right for a tightly edited TV commercial.

If you can get away with it, try to convince the dealership to put a small banner at the bottom of your aerial clip with a phrase like "FAA Compliant 333 Exemption #12345 — Smith's Drone Service." Tell the dealership it will help the aerial pass FAA scrutiny, but it can also serve as advertising for your UAV service.

6. Tower Inspections

Water and cellular towers all have to be inspected on a regular basis for structural integrity, governmental compliance efforts, maintenance reasons, and insurance requirements. In the past some person had to gear up, climb up, and eyeball every foot on all sides of the vertical structure several times a year.

With the advent and acceptance of UAVs, those tower climbers will be replaced by UAVs for a substantial percentage of the time. Every blinking red light you see on a tower at night is an opportunity for repetitive UAV flight income. Drone aerial inspections of towers will be a major market in the coming years.

Drive by the towers and note the phone numbers of the management companies for your cold-call list. I recommend a weekend going out in all four directions for an hour each to make a map of all prospect towers that may need your drone inspection service. From that list, compile a spreadsheet and contact the company that is managing the most towers in your area. One good sales call with them may land you plenty of repeat aerial business deals.

To perform these inspection flights, you will be mainly using your high-resolution still photos rather than HD video. Some tower management companies may require both video and photos, but I would focus on initially gathering closeups from the sides and corners which would total eight pictures from a distance of 20 feet. The repetitions will depend on the vertical coverage your fixed focal lens covers, but leave some overlap.

A typical tower flight of a 150-foot cellular tower will require more than 100 pictures with a higher rate of coverage at the top, where the active components are rigged. You will want to have 16 or more shot angles compared to only 8, that cover the vertical cable runs and support struts. You will want the tower managers to be able to see any cable wear, loose connectors, broken supports, bird and wasp nests, along with any other antenna or cabling anomalies.

Another good selling point is to give them a straight down 90-degree camera pitch shot just over the tower for a current Google Earth view as of that day. That way they can see what the immediate environment surrounding the tower looks like. Also, perform a ground-level pictorial to capture the state of the fencing, gates, road, and UPS system.

7. Events: Weddings, Festivals, and Races

Events can be challenging for numerous reasons due to them being time specific, high profile, and sometimes high risk. Positive gatherings of humanity captured with aerial photography can be very special. As responsible commercial UAV pilots, we will have opportunities to film some extraordinary scenes, but we need to approach those flights with the utmost care and caution.

Remember our number one rule; "fly over no heads," and at the very least minimize your crowd flyover vectors. Burn the battery power if you have to fly a complete circle over trees and buildings to avoid direct human drop zones, just in case you have a power failure or catastrophic equipment malfunction.

If there is law enforcement on site, make contact with them and introduce yourself as a FAA-sanctioned commercial drone pilot to set them at ease. I also contact the local 911 dispatch through a nonemergency phone number to brief them on my flight with time, location, and nature of the coverage. This is to help them if they need to chill out some nervous Nellie who calls 911 about some drone invading the celebration, not knowing the event organizer is paying you to be there.

For weddings and races you have to be quite early most of the time to evaluate the LZ, obstacles, and special client aerial requirements. Plan to arrive at least 30 minutes ahead of the scheduled flight and coordinate with the wedding planner or the race coordinator. Have your observer plan out emergency LZs just in case something goes wrong. Walkie-talkies come in very handy in crowded outdoor venues so bring those along.

With weddings, coordinate with the person running the ceremony so you will be in place when they do the walk and the grand exit. You do not want to be buzzing in the air during the vows when all is very quiet, with the annoying exception of your drone filming angry eyes from the crowd looking at you. Battery consumption, LZ difficulties, and general timing issues can be a real pain with weddings so have patience. Since you will most likely be working with the primary wedding photographer, take his or her cue since that is your client.

The main thing to know about filming races is where the start and finish lines are located. On the race start, try to hover 50 feet up and 500 feet away to comply with your COA. Then as the starting gun goes off, fly in reverse in a rise to capture the running masses from above. This way you can avoid flying over heads but still capture the essence of the competition.

Festivals are not as stressful time wise since the main goal is to capture the crowds having a good time. Same "no head flyover" is in effect, but at a 45-degree angle with your UAV over trees, building roofs, or empty fields you can film the masses with no risk.

Commercial drone photography of public events can be lucrative, fun, and high stress, all at the same time. You will be dealing with dynamic ground conditions, changing time requirements, and high-profile UAV operations. With good planning, close observation, and professional piloting, it can be conducted in a safe and productive manner, but it is all up to you, the pilot in charge.

17 Marketing Your Business

1. Marketing Your Business: Get the Word Out

To run a drone company, you will have to wear many hats and assume numerous responsibilities, as any small-business owner does. Probably the most difficult role is marketing and sales since this will require you to be a bit of an extrovert and to put yourself out there to be seen and heard. You can be the best pilot, with the coolest UAV, making the smoothest aerials, but if nobody buys your product, it is just an expensive hobby.

1.1 Web presence

First, focus on a digital marketing home base which will be your website. Start with GoDaddy.com and pick out a domain name representing your service and location. I selected "CentralTexasDrones.com" to represent my service area. Avoid suffixes like .tv, .biz, or even .net since everybody assumes your company will be located on a .com website.

The website hosting provider needs to have an online website editor for ease of content compilation and formatting. Shy away from complete "free" web hosting since it will surround your content with advertisements, making your site look cheap and a possible malware source.

Weebly.com was the first of its kind to offer a robust online web editor making anyone into a webmaster. With the drag and drop interface, you can quickly set up your website from scratch or pick one of the hundreds of themes as a foundation format. I have numerous websites hosted by Weebly.com for clients of my IT business, with the clients doing the majority of the site updates themselves.

For even more simplistic and seamless website creation, you can use GoDaddy.com for domain registration and website hosting. GoDaddy also has an awesomely easy web editor to utilize, along with a huge library of themes to employ. You will avoid the sometimes confusing step of DNS editing if you have GoDaddy.com take care of both the domain name and the website hosting. Whichever hosting service you choose, plan on spending $10 to $20 a month for enough bandwidth and space to service your website.

1.2 Aerial video hosting

Only have your website text and pictures on your web hosting service, since the space limitation can creep up on you. Start up a free YouTube.com account to house your soon-to-be massive library of HD aerial videos. Since the average size of my aerials are around 400MB, you can see how the storage consumption can add up quickly over months of production.

I have started using the pay-for site Vimeo.com to host my aerials to spare my clients the ads that run on top of my videos hosted at YouTube. I could actually start making a little money from those voluntary YouTube ads, but I would rather have my clients see just their aerials rather than 10 percent of the screen showing a distractive advertisement.

Whichever service you go with, make sure you fill out the description text box along with keywords to help your videos get found and indexed correctly in the search engines. The most important content insert in the description box is the banner stating that you are FAA 333 exempted with the required COA. My standard certification banner that I put on my website, YouTube description boxes, and even within the video itself is "FAA Approved — 333 Exemption #12126."

Both web hosting services at GoDaddy.com and Weebly.com have YouTube video embedding options so you can easily insert the links to your video to have them integrated into your aerial website.

1.3 Social media advertising

Whether you use your current personal Facebook account or start up a separate account devoted to aerial photography, you will have to do it. The more Facebook friends you have viewing your aerial art, the more leads you will receive from people you know. Try to specifically reach out and "friend" realtors, agents, and brokers in your area along with anyone in the construction industry who may give you a call or email inquiring about your drone service.

I am constantly sending out Facebook updates with my latest UAV flight via video links to my YouTube channel. Statistics on my YouTube account show a decent percentage of the views originate from Facebook sources.

Same goes for Twitter. Bite the bullet and create an account if you have not already. Gather as many followers as possible and start feeding them your aerial creations. All this social media advertising takes time and effort, but it is worth it. Marketing is work and it is necessary to generate contacts to make sales.

1.4 Business cards and flyers

Old school marketing still works. You need a good old business card to put in someone's hand when you've been talking about your business. My favorite business card site is Vistaprint.com, since I have used it for numerous marketing projects for the past decade.

Use a nice aerial photo you shot as the background on the card, then insert your contact information and website URL. Get at least 500 cards made in that first batch, since you will be shocked how fast they go once you begin your marketing campaign.

Create a different single 8.5 x 11 colored flyer for each aerial market you are entering. Start with the one for real estate since that will be the easiest to breach, and get off the ground (literally).

You can create these flyers at your home office with a standard inkjet printer. The key is to purchase good glossy photo paper and use the High Quality mode while printing. The result are flyers on photo paper that are thicker and showcase the aerial photos.

1.5 Create market-specific demo videos

Before you head out to a sales call to market your aerial videos, create a single video with all your best pre-market and latest commercial aerials

Low Cost Aerial Photography

HD Aerial Video/Photos Now Available In Washington County and Surrounding Counties

Starting At $300

($300 for up to 100 Acres, $1 additional per acre for greater than 100 Acres)

Deans Consulting (www.CentralTexasDrones.com) is now providing HD Video and High-Resolution Photos of rural properties via low cost small scale aerial drones. Great for Real Estate marketing. Contact FAA Licensed UAV Pilot - John Deans at 979.203.1534

Package Includes: Production level video (2-4 Minutes) in MP4 format and available online, High-Res (14-MegaPixels) photos and mid-level resolution photos all delivered on a portable USB drive and/or DropBox.com. 36"x24" Framed Aerial Photos Also Available.

FAA Licensed Drone Pilot

FAA 333 Exemption: 12126 – Docket FAA-2015-1594 – License #003817933 – UAVs N584DD & N587HD

Flyer

as a portfolio. Do this by starting a new project in PowerDirector 13 and import previous mp4 aerial videos that you have filmed and edited.

String them together, but bring down the audio level to zero so you can overlay longer musical tracks to cover the extended length of the daisy-chained video. For my real estate aerial market, I created a ten-minute collage of numerous scenes and mixed them up to create a serial collage of 60 ten-second scenes from more than a dozen sites. Then I overlaid two long and soothing soundtracks to give it a nice musical background.

Every 30 seconds of this demo video I have a text screen with captions promoting the 1080p resolution, FAA approval — 333 exemption #12126, and the pricing. This is basically a ten-minute running commercial playing in the background as eye candy for the client while I am making my sales pitch.

Create one of these sales demo videos for each aerial market you are pursuing. Start with the most obvious, being the real estate demo. Then create a separate one for construction site aerial status followed by a 2D/3D mapping demo. Preview them extensively and verify there are no glitches or problems before using it as a sales tool.

1.6 Portable demo platform for sales calls

Be ready to close the sale with what you bring to the table. For an initial meeting with just one or two prospective clients, I'll bring a nice laptop equipped with a clear screen running a solid state drive (SSD) that can boot quickly and wake from a sleep mode in seconds. That way right when I sit down with them I open the laptop and start the movie of the demo for that specific market. I will set the volume for the background music so they can get the complete feeling of the production movie, yet not too loud to become distracting to our conversation. Also make sure your presentation laptop has a full charge before the sales call so you will not have to look for an outlet and mess with a power adapter while conducting the meeting.

iPads and tablets also work great for this small audience presentation, but if you use one, have it in a case that can stand it up so you can prop it up in the background and have it playing your top aerial videos while you make the deal.

For real estate sales meetings, you will want to first cold call the broker who owns and runs the company. Then ask for a short ten-minute meeting so you can show the legal and cost-effective way to promote rural properties with FAA-approved UAV/drone aerial photography. With your demo laptop or iPad running your impressive array of aerials, odds are the person will give you a property to fly.

During that initial meeting, I make the offer to return and give a quick group presentation to the resident agents and realtors at that office on the history, legal uses, and future of drone photography in the real estate business. That way, I can get my aerial videos, flyers, and business cards directly to every person at the realty company all at one time.

At large sales meetings like these, I have a 40-inch LED flat-screen Smart TV that fits in the backseat of my vehicle and I can carry it under one arm. It also has the remote control attached to the base with Velcro, and an extension power cord attached to the rear of the TV. The real estate demo video is on a USB thumb drive that is already inserted into the Smart TV's USB port. That way, all I have to do for a large viewing group is to prop the large screen portable TV onto the table, plug it in, and start the video on the USB drive with the remote.

I have had more than a dozen real estate agents mesmerized when viewing my best aerials in full 1080p resolution on a 40-inch LED screen. After giving them a short talk on the correct uses of the UAVs and how we create the deliverable of the production video, and more than 100 photographs for a typical 20-acre rural property, I handed everyone a flyer and my business card. Done deal!

1.7 Join industry groups and give UAV presentations

A sneaky way to penetrate markets to be able to give covert sales talks is to offer generic drone education speeches and presentations to various groups during luncheons and monthly meetings.

Since real estate will most likely be your primary aerial market, consider joining the area group of real estate agents as an affiliate member. This is like an honorary, secondary level for membership to those market-specific groups that represent vendors and other support professionals to that industry. It should only cost you a couple of hundred dollars per year, but it will be worth it due to the networking opportunities that will arise from the membership.

Most of these groups meet monthly or bimonthly with hosted lunches or happy hours. You can volunteer to cover a luncheon or host a meeting, and in return, you get to make a presentation of drone aerial photography, thereby making a pseudo sales call in front of dozens of prospects in one sitting.

Also consider joining the local Rotary Club or Lion's Club for further business networking opportunities. You have to be invited into these organizations, but it will not be too difficult once you start asking around.

1.8 Rent a booth at a trade show or local festival

Having people come to you see your aerials can be an easy way to make sales. Renting a booth at a local trade show or festival is a good way to accomplish getting that traffic to your business. The key here is to make sure you choose an event with high-net-worth attendees. That may sound rough, but it is difficult to sell a $300 aerial video to someone renting a trailer on half an acre of land. Avoid flea markets and any festival where the majority of folks are urban and suburban types. You will want to pay that booth or table fee to a trade show that caters to ranchers and high-value farmers with 20 acres or more.

To set up your aerial photography booth and table, try to have several of the 36 x 24-inch framed aerial photos to hang from the booth supports or canopy roof. That way passersby can get a quick glimpse of what you are peddling. Have a large sign saying something like, "Affordable Aerial Photography" to catch their eyes.

The main attraction to your booth will be that same 40-inch LED Smart TV you make your mobile sales presentations on running the promotional video on a loop. Keep the sound up enough for booth visitors to hear, but not so loud it annoys fellow vendors. Have a large box full of flyers and business cards to hand out to people that so much as pause at your booth.

When people come to a full stop at your table, greet them as a fully FAA-endorsed drone pilot and begin showing them how awesome their estates would look from the air for just a couple of hundred dollars. Try to close sales right then and there by getting their addresses into Google Earth displayed immediately on your iPad. That way you can start planning the flight with them. Next, get their phone numbers and start making flight date plans.

1.9 Direct sales to owners selling rural properties

Sometimes real estate professionals are downright cheap. They want to spend as little as possible on advertising, get the property sold, and collect their percentage as soon as possible. So if the realtor gives you the cold shoulder on a ranch you see for sale and it has great aerial photogenic features, then, using a flyer, contact the owner directly with your sales effort.

Cold calling a resident is very difficult, so try to arrange a happenstance meeting at his or her gate. From there you can mention the "For Sale" sign and bring up your UAV aerial photography offer to help them sell. The sales pitch can be easy if owners are frustrated with the time it has taken to sell the property and they have become impatient with their agents.

Remember, you are making the sale to the property owner so do not get into a situation where the owner expects you to send your aerial bill to the realtor since that won't fly well at all. Be up front on the cost and show the owner the sample deliverables on your laptop. You can also explain to him or her how having this beautiful view from above his or her gorgeous ranch would be a nice keepsake and pictorial memory to have for years to come.

If you are lucky you could make the deal there, fly the property that day, edit the video in the afternoon, and get paid right away.

Take a drive in the country on paved county roads during the weekend and search for real estate signs by nice gates. If there are flyers in a tube by the sign, take one out and look at the acre size and the price. Anything more than $500,000 is a possibility and rural properties valued more than $1,000,000 are a must for you to make a call.

Though cold calls like these are difficult, they are also possible if your timing is right and your information about the state of the sale is accurate. Jump online to that realty webpage remotely from their gate and do some quick research. Then cross-reference a landline number from the address (if they still have a landline). As a last resort, tape a flyer to their "For Sale" sign in the yard and hopefully they will call you.

2. Product Delivery Is Critical

Product delivery is discussed in other chapters, but this is worth repeating: The client has to see the aerial correctly the first time. Since first impressions are everything, the customer must not have difficulty seeing the video or pictures upon delivery. Do not just email a Dropbox or YouTube link or send a USB drive in the mail the first time. There are too many things can go wrong, which can have a negative effect on the client's initial experience of your aerial photography.

Think about it. You convinced a first-time customer to let you fly your drone and film a property for sale, construction site in progress, or a wedding, and all the client ends up seeing upon delivery is a stop-and-start fuzzy video with sound problems. Or worse yet, he or she cannot even read the mp4 file due to the software on his or her computer or TV.

You have to be there in person the first time clients view the aerial production. Make sure you have correctly created the media before you leave your office by testing it on every computer and TV you can access. I verify my mp4 formatted video on my Windows 7, 8, and 10 PCs. Then I test them on my iMac, iPhone, iPad, and Android tablet.

Yes, I know most people do not have all those platforms, but you get the idea of using all devices you either outright own or at least have access to. You may have friends and family that have different OS-based devices you could test your mp4s, AVIs, or even WMV formatted videos. Realistically speaking, you need to do it only once as long as your production process within PowerDirector 13 does not change. If you stay consistent with the output formatting specifics, then repeat testing will not be required.

As long as your aerial media is reliable, all you have to do is play IT guy and make your video view nicely on clients' displays. Remember, take your laptop or Smart TV with you to the clients' site during the first aerial media delivery just in case all their computers, laptops, and TVs are junk. That way at least they will be able to see how good your product looks on your equipment. Then you can suggest what they should purchase or upgrade to see future aerials on a rich display of their own.

Try to shy away from just sending them links to YouTube or Vimeo since it all comes down to Internet service provider performance. One day the clients may view an aerial on YouTube you performed for them last month and it played in high definition without buffering and looked great. Then the next month's construction aerial status they received from you came when a high utilization their DSL connection was going on and the video played in less than HD with stops and starts. The end result was your aerial looked bad.

In other words, you are taking responsibility for how and what they see when your product is delivered and viewed for the first time. Try to eliminate any computer, graphics, network, or media problems that may have a negative effect on the enjoyment and satisfaction your client has with your aerial creations.

3. Groom Clients to Help Sell Your Services

It can be said that a happy client equals more sales and income. Since aerial photography with UAVs is such a new industry, we need very satisfied customers to help spread the word so we can expand our markets. The neat thing about what we do is that our clients can help show other people how beautiful our products are.

Immediately after blowing away your clients with the awesome aerial your UAV captured, draft them into helping you sell more to their peers. Come with half a dozen USB drives with copies of their video and pictures so they can give those to their coworkers, friends, and even family members. Accompany those USB drives with your flyers and business cards.

Actually ask them who else you can contact that they know that will benefit from your aerial service. This the best opportunity you will have; while they are happy with you and your abilities. Get ready to write down names, numbers, and email addresses.

Happiness only lasts a little while, so seize the moment and deploy your satisfied clients as your new salespeople. Just like in the initial sales presentation, offer to return to deliver a generic talk to other employees or business partners about the current state of affairs with drones and the future of UAV photography in their businesses.

18 Pro bono Test Flights

Now that you have every aerial entity and editing process ready for deployment to earn out in the field, first try the complete process free of charge for family and friends. Many of us know people or have relatives to whom we can offer our pro bono aerial services for testing purposes. This way, you can double check your gear, sharpen your piloting skills, and find any flaws now while it is much less stressful to resolve.

Try to perform one, if not two, of these pro bono aerial projects in each specific market you are planning on pursuing. This could add up to more than a dozen free flights over numerous weeks, without making any money. The effort will be worth it when you do find little problems that are easy to resolve and even more valuable when you find showstopper difficulties that would have made you look like an amateur in front of a client.

These nonprofit flights can be made during the waiting period for your FAA 333 exemption to get approved or your UAV license to arrive once the feds get that rolling. You will have even more time while awaiting the November registration numbers from the FAA's Oklahoma office.

19 Pricing Your UAV Services and Products

Drone aerial photography will not make you a millionaire. In fact, the first few months will be a slow, part-time job. Pricing your work can be tricky, since this is such a new industry. Rogue droners charge anywhere from $100 to $1,000, depending on what they are filming and who they are filming it for, with the risk of a $10,000 fine from the FAA if they get caught.

We 333 exemption FAA-approved UAV companies with registered birds and licensed pilots can eliminate the risks of fines from our worried minds. That said, we still have to fairly charge for all the time and effort it took attaining those exemptions, certifications, registrations, and qualifications.

Every service and product has a value. Your time also has a worth assigned to it based on your current employment status. If you are a white collar professional making $300,000 a year, then flying drones for a few hundred dollars a flight may not be worth your time. Other factors in play are family responsibilities, outstanding debts, and your current career development. This may not be the time to become a professional flying camera pilot.

If your situation does fit the bill for the UAV business, then do your market research on the pricing points for the aerial market you are going to be pursuing. The goal is to charge as high of a price possible that the demand will accept and your competition allows.

1. What's the Competition Charging?

Here you will have to compare apples to apples. Try to find similar services that are currently serving your area or at least one in your region or state. Do not stray too far by comparing your proposed flight prices in your home town of Brenham, Texas, to overpriced UAV guys flying Sonoma, California, ranch properties.

The key to pricing is knowing what your regional market will pay. The first route to seek this information is to web search for your competition. Find a company that is active in the aerial market you are entering, and find out what it charges and how it scales pricing. Most times it will be on the website on a pricing or products page. If not, simply call and politely inquire.

After you have collected pricing information from your competitors, observe what their deliverables look like. Are yours better, the same, or lower quality than theirs? Hopefully you will be able to deliver a higher quality aerial video and photograph array at a comparable or even lower price.

2. Add-ons Enhance Your Price

Try to deliver a comprehensive and thorough aerial product. Provide a smooth, cleanly edited, and powerful production video. Include as many high-resolution photos independently shot outside of the video along with twice, if not three times, as many stills taken from the video. It should add up to more than 100 aerial photographs.

Another addition you can include if competition is especially fierce, is a 2D map photo as an extra to close the deal and make the client happy. This can be done easily if the property is less than 20 acres and you have some experience performing mapping runs.

One last extra service you can perform for real estate deals is to photograph the inside of the home, and provide a voiceover summary of the property features. This requires extra equipment and talent, but if you can pull it off, it will be worth it.

3. Ramp up Pricing

If you have no idea how to price your aerial services, then try a slow ramp-up in charges. You can start by asking for $200 for a simple video and 50 photos for the first client. Hopefully the first opportunity will not be the biggest in your area. Start small, get your process bugs worked out, and then move up to more challenging clients.

After you get a few projects successfully flown, edited, delivered, and paid, you can increase your charge from $200 to $250. As the months pass, along with your successful project completions, your prices can escalate until you hit the ceiling of that products' worth in the specific region during that time. You may max out at $300 or even $500 depending on your skill set, the heat of the real estate or construction market, and the part of the country in which you are working.

20 Running a Business: The Paperwork and More

1. Run Your Business for the Long Term

What you are doing is no longer a hobby. It is a serious money-making operation and it must be run as an authentic company. Accounting and governmental compliance is a royal pain but is absolutely necessary for longevity and financial survival.

2. Incorporate First

As we briefly discussed in the business plan, you will need to incorporate your business as soon as possible for liability and tax purposes. Check with a Certified Public Accountant (CPA) in your area to help you pick whether your business will be an LLC, S-Corp, or an LLP. Your state law will likely determine which type of corporation you can file for, so start educating yourself on that subject.

Whichever flavor of corporate filing you choose, get it done before you start flying for money. The primary reason you are going through this additional paperwork jungle is to help shelter your personal assets from liability. Remember, bad stuff does happen and drones can fall from the sky.

Separating your business and personal expenses is critical also for tax reasons. Having a solid corporation set up properly will help contain the financials within that entity and have less of a negative effect on your personal taxes. Trying to run your aerial business under a DBA (Doing Business As) is just not sufficient. You will need a proper state-endorsed corporation for not only taxes and liability issues, but also to help when you file for your FAA 333 exemption.

3. Set up QuickBooks Accounting

A real business has to have a real accounting system. I strongly suggest purchasing the $200 QuickBooks accounting software from Intuit. For more than ten years now, I have run my independent companies with QuickBooks and have had very good success. We run payroll, state and government reporting, and year-end taxes for local, state, and federal entities.

Every transaction needs to be entered into your QuickBooks software. That includes purchases of UAVs, batteries, and even USB drives. When you deliver an aerial it will be accompanied with an invoice generated from QuickBooks. When you write a check for your new Phantom 3 Professional, it will be entered into your QuickBooks.

Set up a new business account at your bank and have QuickBooks track each and every deposit and debit from the start. I cannot stress enough how critical it is to keep personal and business expenditures separate.

4. Pay Company Taxes

Another accounting practice you should follow is to claim all income to your new aerial business. Even if the client insists on paying you in cash, take those paper dollars straight to the bank and deposit them. Show all incoming payments in QuickBooks and run them through the new business bank account you set up. It will make your quarterly and year-end reports authentic and informative.

One of the main reasons for all this accounting talk is to keep you and your business out of IRS trouble. So many small businesses fail due to problems with the nation's Internal Revenue Service because of sloppy accounting practices, hiding income, or not paying employment taxes. Ask your accountant to help you set up the QuickBooks company file with the proper payroll configuration so you can pay yourself properly and make those wonderful quarterly payroll tax payments to the IRS.

Check with your state and local taxing authorities to see if you need to charge sales tax on each aerial flight. Every state is different, and some may be quite lenient on your new service. Odds are if local professional photographers have to charge state or local sales tax, then you will also. It is much better to know and implement this in your invoices now than learn about it months or even years later owing thousands in back-sales taxes after an audit.

None of this accounting stuff is fun and it can be a real pain to deal with on a monthly basis. But if you want to make a living in the long term with a flying camera business that is fun, take the time to set up everything financially so you can do your dream job.

5. Start Slow, Then Grow Your UAV Fleet

Most likely, you will start your drone company up with a single bird. You may start out simple and low cost with a used Phantom 2 Vision Plus, or have enough initial investment for a DJI Phantom 3 Professional or even an Inspire 1. With only a single UAV in your hangar, you will have to be very cautious with your drone, and not take any unnecessary risks.

This means to only fly the UAV for money and not for fun when your buddies swing by and have had a few beers. My company drones only take off when the money meter is running. In other words, I am not risking my primary revenue-generating aircraft on any extracurricular activities that could go wrong and shut down my earning capacity.

5.1 Always have a backup bird

Remember that two is one and one is none. Bad things can happen quickly when manmade objects are in the air. Again, Murphy's Law is always in effect, and that goes double with UAVs. Keep in mind that we rely on numerous sensors, motors, propellers, batteries, power supplies, computers, network transmitters, space satellites, compatible operating systems, and software applications to work simultaneously to keep our birds stable in the air. Even if a single UAV component fails, whether in the air or on the ground, our drone could come crashing down.

This is why I strongly recommend you purchase a second UAV to serve as a backup in case of a temporary breakdown or complete loss of your primary drone. You can do that by saving the income from the first few flights, then purchasing a second bird to add to your fleet.

Drone with Recovery Message

Whenever I am going to fly an important first-time site or make a trip to multiple sites in a single day, I take both birds; in case something goes wrong with the first UAV I can deploy my backup and keep earning.

Along with the FAA required registration November number labeled on your UAV, also place your cell phone number on the aircraft with the word "Reward" on it to help it get returned in case of a flyaway. I also name my UAVs with labels on the fuselage and the remote just to keep them straight and orderly. My first two Phantom 2 Vision Pluses were named Jake and Elwood, from *The Blues Brothers*.

5.2 Enhance and improve your birds

With drone technology moving so fast and UAV pioneering companies like DJI and US Robotics making leaps and bounds in the marketplace, you as a drone company owner will have to keep up with growing expectations.

Clients read about new enhancements and expect your birds to have them when they hire your company for aerial services. Very soon it will just not be enough to show up at real estate flight site and film it with a Phantom 2 Vision Plus using the standard integrated HD camera. More and more clients will require you to shoot in the best quality and highest resolution, which is now the 4K level of cameras.

Other higher levels of performance will also be expected now that UAVs can fly with GPS-like precision indoors via new bat-like sonar sensors. Capabilities such as longer flight times, better camera controls, and automated flight abilities will also be the call of the day.

Most likely, and sooner than you may think, the purchase of a new Phantom 3 Professional or Inspire 1 will be on your company's to-do list just to meet the growing customer expectations.

Bottom line: Immediately start putting a portion of your aerial company's income toward improving your fleet of UAVs. Be sure you have a spare for business continuity in the face of adversity. After that has been accomplished, keep an eye on the latest drone improvements and capabilities so you can capitalize on that with the purchase of an upgraded and enhanced model of UAV.

5.3 Stick with the Phantom 3 Professional recommendation

As the owner of two DJI Phantom 2 Vision Plus birds, I have been waiting for the next big upgrade in both UAV capabilities, camera quality, and piloting control enhancements. I was tempted with the Inspire 1 which came out in late 2014, but at twice the price for the aircraft and batteries compared to the Phantom 2 Vision Plus, I decided to wait a bit.

The summer of 2015 brought the Phantom 3 in the Standard, Advanced, and Professional versions. Since the price of the fully equipped Professional was nearly identical ($1,300) to the Phantom 2 Vision Plus original price I paid in 2014, I chose the top model of the Phantom 3 to become my next generation UAV for Central Texas Drones.

5.3i Hardware improvements

The power plant has been beefed up providing stronger lift capabilities. The Phantom 3 batteries look the same, but take more amps to charge them up from a new charger, and deliver longer flight time to the more powerful rotor motors. I was bummed about the incompatibility with the batteries since I was hoping to use the six Phantom 2 Vision Plus batteries I had built up over the past year to use for the new Phantom 3. The new power adapter charges up both the primary Phantom 3 battery and the remote controller, but not at the same time; not recommended anyway.

The Chinese company DJI decided to enable the Phantom 3s to utilize both USA and Russian GPS satellites for faster acquisition and better position precision. I noticed that right away when my new Phantom 3 Professional was able to sync to 13 satellites in less than a minute.

The other big piloting improvement of the bird is the new optical flow and ultrasonic sensors to enable the bird to fly better without GPS coverage. Basically, the Phantom 3 becomes a bat indoors which opens up new marketing possibilities.

Flying distance was more than doubled by both the new dual antenna system on the remote controller coupled with the Lightbridge FPV (First Person View) technology in the Phantom 3 drone. This means you get nearly full HD video on your display with little delay at distances up to one and a half miles.

The new RC unit is completely redesigned with numerous function buttons, longer range dual antenna, and a built-in rechargeable battery. The remote controller turns on just like the Phantom does and has similar power charge indicator lights. It's got a nice thumb wheel to control the camera pitch and separate buttons to take videos and snapshots. Gone is the external Wi-Fi extender that needed recharging. My favorite new button is the "Go Home" feature, which can also be pushed again to regain control of the bird once it returns within range.

Finally, the Phantom 3 Professional's new 4K camera shooting at 30fps is crystal clear. The file sizes are huge at 8MB per second so hopefully the clarity is worth the disk space. Filming with the Phantom 3 Professional's 4K (3840 x 2160 pixels) resolution at 30 frames per second (FPS) will create mp4s that are four times larger than the same movie files captured by your Phantom 2 Vision Plus at Full HD of (1920 x 1080 pixels) at the same frame rate.

Then you have to consider on which graphics display you are, and more importantly your client is, going to view that ultra-high resolution video appropriately. As of late 2015, very few affordable 4K computer monitors, tablets, and Smart TVs are available. Just to prepare for the near future, it may be prudent to film in 4K, but get ready to buy numerous terabyte external drives to store the giant video files.

Another issue to consider is the strength of the video-editing workstation and the ability for the editing software to handle a dozen or more 30-second video clips that are 240MBs each. Even with my Intel I-5 processor, 16GB of RAM and working on a solid state drive, my PowerDirector video-editing software dragged to a crawl and the previewing was very choppy when I tested compiling an aerial filmed in 4K resolution.

Bottom line: I'm only going to film special projects directly intended to be displayed on near-future 4K devices and stick with the full HD 1080p resolution for the majority of my projects. That will change when a significant percentage of computer monitors, tablets, and Smart TVs display 4K resolution.

5.3ii DJI Go App

The biggest transition you will encounter in upgrading from the Phantom 2 to the Phantom 3 will not be the piloting or the bird itself, but rather migrating to the smartphone/tablet app the runs everything. Changing over from the DJI Vision to the DJI Go app is massive since the complete user interface has been redesigned and significantly enhanced.

I highly recommend relearning all the basic operations, start up procedures, and emergency actions when first piloting the Phantom 3 under the DJI Go applications over several days and numerous flights before taking it out into the field to earn.

5.3iii Tablet compatibilities

After purchasing my Phantom 3, I quickly learned that the iPad I had interfacing with my Phantom 2 Vision Plus was not strong enough; therefore, not compatible with the required DJI Go app. Luckily the Verizon VK 700 I was testing worked with the Go app, which spared me from purchasing the new iPad 3. The downside of that tablet change was I had to get just as handy with the Android OS as I was with the Apple iOS on the iPad.

If you are pondering which tablet to purchase to run your Phantom 3, you will need to check the DJI site for specific makes and models that meet the strict compatibility requirements. The thing you do not want to do is buy an expensive ten-inch tablet with a cellular contract for a year to find out later that it is not compatible with your DJI Go app.

5.3iv Phantom and RC Firmware Upgrades

Since most DJI products will have been sitting on shelves, on pallets, or inside shipping containers for numerous weeks, if not months, both your bird and the remote controller will be out of date on the firmware.

I assure you that the version of the DJI Go app will require the latest revision of the firmware for both the UAV and the RC. The best thing to do is to immediately follow the upgrade procedure recommended by DJI utilizing the Micro SD card with the downloaded .bin firmware file to be uploaded into the Phantom on power up.

Upgrading the Phantom 3 can be a test of your patience. Watch some YouTube videos on it so you will not be tempted to reboot the bird thinking it is hung up during long processes of the firmware upgrade. The minimum time is 25 minutes and mine took nearly an hour. It is just as critical to upgrade the remote controller to keep harmony between the firmware revisions.

5.3v Intelligent flight modes

The long-awaited advanced flight capabilities were finally delivered via the firmware upgrade of 1.4.10 to the Phantom 3 and the latest version of the Go app by DJI in September of 2015. The primary enhancement was the implementation of the Intelligent Navigation Modes (INM).

Intelligent Navigation Mode (INM)

This is activated by moving the three position switch on the right from P to F which will bring up the INM menu. Course Lock is a great flight feature enabling the pilot to establish the forward vector and while moving forward, he or she can move the right and left turn stick panning the scene without changing the drone's course vector. This can make for some awesome fly-by views.

The next mode is the Home Lock which can save you from being lost in the UAV's directional perspective. With Home Lock enabled, you can simply pull back on the forward/reverse stick and the bird will begin flying in the direction of home no matter what the current location is or which direction the drone was pointing.

My favorite INM mode is the Point of Interest which makes any pilot fly like a pro around a center point. This is the big one enabling us to fly around a ranch house or construction pad in a perfectly smooth circular flight with the camera pointed consistently toward the inside target. Before this INM feature, it was all up to the steady hands and fingers of the pilot, and it was very difficult to maintain a flawless arc around the point without any abrupt movements, turns, or jerks.

In Follow Me mode, you can walk, run, or ride in a vehicle and the Phantom will follow you. This can be helpful in numerous types of event coverage projects. Certain minimum distances and elevations are required during startup for safety reasons. Also, just be careful not to travel under raised obstacles like trees, power lines, or bridges.

Finally, the Waypoint mode is what you will now use for the mapping projects. Gone are the days in the DJI Vision app where you premapped the waypoints and then sent the bird off to fly your preflight course. Now, you have to actually fly the grid first and set the waypoints in real time. Only after you have flown the pattern and set each waypoint, can you put the bird aloft to fly it in the proper speed and camera recording mode.

Though the Phantom 2 Vision Plus and the Phantom 3 may look quite alike, the new user interface of the DJI Go app along with the numerous advanced flight capabilities offered in the Intelligent Navigation Mode suite will send you back to UAV flight school for many flights to learn how to best implement them in your aerial photography business.

6. Other Administrative Tasks

Some of the other issues that come up when you run a business have been covered in other chapters, such as your initial investments (Chapter 6), and initial staffing (see Chapter 11 where observers are discussed).

Running a drone business also means you'll have licensing issues to keep up with, for flying and for the actual business (your city likely requires you to have a business license). You'll want insurance (drone insurance was touched on in Chapter 9, but also other equipment, vehicle, and business insurance). It may seem obvious, but there are other bills that come with running a business, such as your cell phone bill, and marketing costs as discussed in Chapter 17 for your business cards, website, and anything else you do.

It is a good idea to list everything you can think of that will be either an administrative task or a business bill so that you can be on top of things once your business gets going.

21 Aerial Road Trips: Mobile Recharging and Editing

Not every billable UAV project will be just down the street from your home. In fact, the more business you line up, the farther you will have to travel. Since time is money, we want to be as efficient as possible in our distance logistics, so planning is everything. There are some effective techniques you can utilize to be a dollar-wise drone road warrior.

1. Planning a Multiple Flight-site Trip

There will be times when projects require an hour or more in travel. Some may even require a plane ticket, but those will be rare. Most aerial gigs will be in your region usually within 100 miles. This natural distance limitation will be due to clients not wanting to pay travel costs. Plus with the growing UAV talent pool, your competition will cover those long-distance clients.

So when your flight project list has one located an hour to the south, a few more to the west, and a final one just north of that, try to group them on the same day, if possible. That way you can make a big sweeping road trip knocking all four out from late morning until just before sunset. This beats making four separate trips on four different days that could bring bad weather. If you have a nice day with little wind and clear skies, then it's time to fly as many projects as possible if the distance legs in between are just right.

Multiple flights in one day can be intense, profitable, and logistically difficult. You will have to order the flights based on availability of an observer, site accessibility, and specific activity required at the site like those requested at construction projects. The light of specific times of the day can play a role for your trip planning also. I would have my more scenic sites scheduled for late in the day, since they require the best lighting.

Also remember to plan to spend some time on the phone with the FAA filing your NOTAMs. Hopefully, you will get a patient flight planner on the phone when you say you need to file five flights for one day. My experience is I usually am on the phone with the FAA 10 to 15 minutes for each flight site NOTAM filed. That means you may burn at least an hour on the phone a couple of days before the UAV road trip.

2. Mobile Battery Management

Hitting numerous flight sites in a single day can challenge your battery charging capabilities. Having an in-car DC/AC inverter for the 110-Volt chargers is mandatory. Three to four sites would likely take 9 to 12 fully charged Phantom batteries, and having that many in your active inventory would be quite expensive. Since most of us stock 4 to 6, we would have to recharge them in between each flight site.

I recommend having a high-quality DC/AC inverter directly wired to your vehicle's electrical panel and installed under the dash near the glove compartment on the passenger side. The high-watt inverter should have multiple 110-Volt power sockets along with a few USB power ports. That way you can also recharge your RC controller, iPad or Android Tablet, Wi-Fi range extender, and your smartphone while you are driving. If you are flying in a secure area, keep the car running and charging your Phantom batteries so you can stretch those four batteries across a ten-battery flight array throughout the day.

Recently, I performed four flights at four different sites with a combined driving distance of nearly 200 miles. Since the sites were at least 45 minutes apart, I was able to charge the batteries used at each previous site. By the end of the day I had more than 10GB of HD video and hundreds of pictures.

3. Make Backups between Sites

During those multi-flight days, make sure you start with an empty Micro SD card so all the video and pictures will fit. I also take my laptop so I can back up the accumulated media after each site to my internal laptop drive to protect my work. This saves me from losing the aerial footage of my labor if I lose my bird on the final sortie at the last site. It would be bad enough to lose a drone, but to suffer the loss of all the work done that day up to that point would be horrendous.

Another way to have on-the-road redundancy is to take multiple Micro-SD cards with you and swap them out at each site. You will have to stay organized by having a spreadsheet noting the flight sites with times and scene location specifics. It is also a good idea to copy the videos and photos in the site-named folders on your backup drive.

4. Onsite Editing and Production

You may have an opportunity to hit a site that is far away only once, and you will not want to drive back there the next day after editing and producing the video that night. One solution is to have all the production software on a powerful laptop and perform those after-flight operations in the car while parked safely off the highway. Finding a good Starbucks or restaurant to settle down and produce the aerial deliverable is a better idea and it is much more comfortable than working in your vehicle.

Since we always want to present the initial aerial in person to make sure the client sees the video clearly and enjoys it, you may want to fly the site, produce near the site after the bird is down, and present your deliverables right then and there. Another benefit of this on-the-road production process is that you can get paid the same day if you request payment on delivery.

Time may limit multiple fly-and-produce projects in a single day, but they are an option if you have the portable equipment to accomplish the work with the same amount of editing quality as performed in your home office.

5. Vehicle Security and Safety

During multiple flight-site trips, you will be very busy while driving and flying. When you are behind the wheel, do you best to focus on the road and not fiddle with rechargers, monitor the weather apps, and download video from the UAV. Pull over and park the vehicle way off the highway to change batteries on your charger or manipulate other drone equipment.

Maintain your situational awareness while sitting in your parked car and do not let potential bad guys sneak up on you. Remember to lock your doors and keep an eye open for dangers. When you are out of the vehicle make sure to hide your expensive gear by stowing it in the trunk or covering it with a blanket.

If a car thief catches sight of a laptop on the backseat, it may be just enough to entice him or her to break into your unoccupied car while you are eating at a restaurant in between flights. Once in your vehicle, what is to stop a thief from taking all of your UAV equipment and wiping your livelihood away with a single theft? Lock up, hide your stuff, keep an eye out for the wolf, and have your observer keep watch too.

22 Offer UAV Consulting and Configuration Services

Since we are the commercial drone pioneers emerging from the early rogue pilots and later becoming legitimate "333 exemption" holders, it will be up to us to teach and train newcomers. Already, companies are purchasing their own drone assets and are either clueless on how to implement them in the businesses, or pilotless to legally operate above ground.

It will be up to us to fill those voids and seek emerging opportunities for UAV consulting and drone configuration. Since college degrees for drone specialists aren't available, real world experience is everything, at least for now. We need to seize today's opportunity and offer our acquired expertise in UAV operations.

1. UAV Education and Consulting

There are companies today watching how Amazon implements its drone-based delivery service and contemplating how their companies can utilize UAV capabilities. Put together a PowerPoint presentation on commercial UAVs and customize a version to that company's market focus. This will enable you to charge for presentations and industry briefings, since you will have the credibility and real world experience as a commercial drone pilot.

Companies that do purchase UAVs will have to undergo flight training or even hire certified pilots to fulfill their FAA 333 exemption requirements. Even after the formal FAA UAS licensing process is in place, the primary need for UAV personnel will be from the pool of those flying today. Offer your experience, but charge handsomely for it.

2. Outsourced UAV Pilots

The initial workload of drone pilots will not justify full-time positions for most small- to mid-sized corporations. They will opt to hire part-time pilots to fly their new UAVs, and the best resource will be seasoned commercial droners with serious stick time and good standings under the FAA who are 333 exemption holders.

Watch the "help wanted" section of the newspaper or classifieds websites for "UAV pilots wanted" (coming soon!). It may be worth taking a job like that, even if it is part time, just to get time on higher-end birds. Think of it as building your drone résumé and logging more flight hours.

As a small-business owner, it would be difficult to take a full-time job as a corporate UAV pilot, but filling in a few hours a week could enable you to make extraordinary contacts, which may lead to future sales.

23 Going Forward

1. Stay ahead of UAV Technology Offerings

The UAV industry began its explosion in 2014 much like the computer world did when I got my start in IT back in the early 1980s. As commercial drone pilots, we need to offer our services based on the leading edge of this technology.

2. Push It to the Safe Edge

We can experiment and test with new UAV systems that skate the bleeding edge, but we need to standardize on the more stable leading edge. In other words, operate your birds within the operating parameters when on paying jobs. The primary reason for this precaution is safety. The second is to keep your drones operational to earn as much as possible while they are airworthy.

One idea is to have a third bird that is reserved for testing new features and robust customizations. This can be your development drone to experiment with longer range antenna, more aggressive props, and tweaked-out firmware providing more in-flight options.

3. Stay Informed to Keep up Your Expertise

Sign up for every drone newsletter and UAV blog post you can find. Configure Google News Alerts like I did for all of 2015. It sends me a daily email packed with every news article from around the world having anything to do with the words "drone," "UAV," "UAS," etc.

Join local and regional drone user groups, flight teams, and related organizations that physically get together to fly UAVs and support each other's airborne adventures. Learn from other pilots about how and what they are flying. Find out what problems they have encountered and how they resolved those issues.

Try your best to attend at least one UAV trade show. When you have the funds and especially when one is held near you, bite the bullet and spend the money to attend. It will be a couple of days of nothing but drone hardware, software, accessories, and learning opportunities. The networking potential alone is well worth the time and costs to attend.

Only there can you walk the aisles of all the vendors, UAV builders, drone enthusiasts, and industry experts. Web surfing drone information on the Internet is one thing, but getting your hands on new technology, seeing the latest UAV models, and hearing about the next generation of drones will keep you on the leading edge of remote aerial technology.

The photography industry is also represented at UAV trade shows which helps you cover the aerial camerawork that is critical for your business. Try to attend at many talks and presentations as possible taking notes or recording the audio for future references.

4. Develop Vendor Relationships

Get to know the main players at the UAV trade shows and the local hobby shop guys that may end up supporting you one day after your drone goes down. If you can get on a first-name basis with these UAV specialists, they tend to offer you better support and insider information of new features coming out and even what to avoid flying that's on the market now.

My favorite drone shop support team is Nathan and Eric at UAVDirect.com, which is north of Austin,Texas. Numerous times, I have called them to ask questions about firmware and other upgrade ideas for my Phantoms. They have been very helpful and I am planning to have them build my Agricultural-Drone with NDVI (Normalized Difference Vegetation Index) to expand into that market in early 2016.

5. We Are the Pioneers

As 2015 wraps up, the UAV industry is on fire with extraordinary activity, growth, and technological advancement. We are riding the tide of the drone tsunami just trying to keep up with the latest drone model and the newest firmware version. Thousands of the rogue pilots that were charging for flights outside of the FAA rules are now coming into the fold by filing for their 333 exemptions and migrating into certified commercial UAV pilots with valid COAs.

Though DJI and 3D Robotics are leading the consumer drone industry, many more companies will emerge to challenge their technology, aerial capabilities, and eventually their market share. That free-market competition from around the world will only benefit us commercial droners by providing the best hardware, software, and aerial configurations possible.

Some think that all this drone stuff is just a novelty which will soon wear thin and fade away just as fast as it flew in. I remember those same type of pessimists saying that about the Internet 20 years ago. One possible similarity may be the boom we saw in the late 1990s with all the .com businesses that sprang up overnight only to go bust in the tech crash of 2000.

Though there will probably be some thinning out of the latecomers to the UAV markets, the unmanned aerial world is here to stay and we have a strong foothold into it as commercial UAV operators.

Who of us would have thought five years ago that it is now possible to earn a living by flying an RC aircraft with a camera attached to the bottom of it? Even stranger jobs for UAVs are being implemented every day now. We have "shark drones" that patrol beaches on both the east and west coast lines keeping a watchful eye out for water- bound predators. Amazon plans to be delivering packages via UAVs by 2017 on a regular basis. Cardiac defibrillator-equipped drones will be just a button push on a smartphone away in metropolitan areas soon.[1]

As the Internet technology market did in 2000, there will be swing downs in the opposite of the high flying industry peaks we are experiencing now, so be prepared for it. After the FAA commercial UAV rules are

1 "Ambulance drone delivers help to heart attack victims," Michelle Starr, CNET, accessed November 2015. www.cnet.com/news/ambulance-drone-delivers-help-to-heart-attack-victims

finalized and implemented, there will be a massive influx of want-to-be commercial droners that will cut prices and flood our industry.

That will be the competition battle between the 333 exemption holders and the new group who only had to take an online FAA test and pay the $100 commercial drone pilot's license fee. Expect your prices to be challenged by the next generation of commercial droners and even that you will lose some business to these lower-cost aerial services that will swamp our marketplace.

The key to your survival will be the quality of your products and your existing customer relationships. What is true for most every business is that being able to keep the customer happy during every stage of a market is a rollercoaster ride.

Keep up on the rapid technology changes and upgrades the UAV industry is bound to go through, and maintain excellence in your aerial photography deliverables and client relations. Keep those positive aspects airborne and you will survive the sure to come ups and downs of the commercial UAV market.

6. Wrapping up

There are two big first impressions I recall that got me hooked on becoming a commercial drone pilot. The first was when I lifted my Phantom off the ground the day it arrived and was simply amazed at how it just hovered without me having to maintain the controls. The next "wow" moment came just after I downloaded the HD video taken of my rural property and played it back on my computer screen.

Viewing the spectacular scene in high resolution took my breath away. I was looking through a bird's eye and seeing what they see every day, but now I was the one flying. "People need to see this. They need to have their properties photographed and videoed in this aerial format to capture the beauty of the land from above," I thought.

To enjoy working, we all need to do something about which we are passionate. Something we can wake up in the morning and look forward to doing during the day. We need mental stimulation to keep us from getting bored with our day-to-day chores and responsibilities.

After numerous years in Information Technology looking for another line of work that both interested and challenged me, I truly believe I have found that new career as a commercial drone pilot and aerial photography business owner.

I also believe that there are tens of thousands of others who can also pursue this satisfying and fascinating endeavor from all over the country, and millions more throughout the world. We can now have one of the coolest jobs on the planet thanks to the flying camera.

Keep in touch: Check out my website at www.centraltexasdrones.com, and my Central Texas Drones YouTube.com channel at www.youtube.com/channel/UCpi-3znUXVBIAJcXDS8bEvQ; there are some examples of two primary aerial market projects that reflect numerous aspects of what we have discussed in this book. The download kit included with this book contains samples of my work as well.

Download Kit

Please enter the URL you see in the box below into your computer web browser to access and download the kit.

www.self-counsel.com/updates/dronepilot/16kit.htm

The download kit includes:

- Sample UAV videos
- Web resources to help you as you launch your drone business